Lecture Notes in Electrical Engineering   1006

The book series *Lecture Notes in Electrical Engineering* (LNEE) publishes the latest developments in Electrical Engineering—quickly, informally and in high quality. While original research reported in proceedings and monographs has traditionally formed the core of LNEE, we also encourage authors to submit books devoted to supporting student education and professional training in the various fields and applications areas of electrical engineering. The series cover classical and emerging topics concerning:

- Communication Engineering, Information Theory and Networks
- Electronics Engineering and Microelectronics
- Signal, Image and Speech Processing
- Wireless and Mobile Communication
- Circuits and Systems
- Energy Systems, Power Electronics and Electrical Machines
- Electro-optical Engineering
- Instrumentation Engineering
- Avionics Engineering
- Control Systems
- Internet-of-Things and Cybersecurity
- Biomedical Devices, MEMS and NEMS

For general information about this book series, comments or suggestions, please contact leontina.dicecco@springer.com.

To submit a proposal or request further information, please contact the Publishing Editor in your country:

China

Jasmine Dou, Editor (jasmine.dou@springer.com)

India, Japan, Rest of Asia

Swati Meherishi, Editorial Director (Swati.Meherishi@springer.com)

Southeast Asia, Australia, New Zealand

Ramesh Nath Premnath, Editor (ramesh.premnath@springernature.com)

USA, Canada

Michael Luby, Senior Editor (michael.luby@springer.com)

All other Countries

Leontina Di Cecco, Senior Editor (leontina.dicecco@springer.com)

** This series is indexed by EI Compendex and Scopus databases. **

Oleg Gusikhin · Kurosh Madani · Henk Nijmeijer
Editors

Informatics in Control, Automation and Robotics

18th International Conference, ICINCO 2021
Lieusaint - Paris, France, July 6–8, 2021,
Revised Selected Papers

 Springer

Editors
Oleg Gusikhin
Ford Research and Advanced Engineering
Dearborn, MI, USA

Kurosh Madani
University of Paris-EST Créteil (UPEC)
Lieusaint, France

Henk Nijmeijer
Department of Mechanical Engineering
Eindhoven University of Technology
Eindhoven, Noord-Brabant, The Netherlands

ISSN 1876-1100   ISSN 1876-1119 (electronic)
Lecture Notes in Electrical Engineering
ISBN 978-3-031-26476-4   ISBN 978-3-031-26474-0 (eBook)
https://doi.org/10.1007/978-3-031-26474-0

This Springer imprint is published by the registered company Springer Nature Switzerland AG
The registered company address is: Gewerbestrasse 11, 6330 Cham, Switzerland

Preface

The present book includes extended and revised versions of a set of selected papers from the 18th International Conference on Informatics in Control, Automation and Robotics (ICINCO 2021), that was exceptionally held as an online event, due to COVID-19, from 6–8 July 2021.

ICINCO 2021 received 121 paper submissions from 38 countries, of which 8% were included in this book.

The papers are selected by the event chairs, and their selection is based on a number of criteria that include the classifications and comments provided by the program committee members, the session chairs' assessment and also the program chairs' global view of all papers from the technical program. The authors of selected papers have then been invited to submit a revised and extended version of their papers having at least 30% innovative material.

The purpose of the International Conference on Informatics in Control, Automation and Robotics (ICINCO) has been to bring together researchers, engineers and practitioners interested in the application of Informatics to Control, Automation and Robotics. Four simultaneous tracks have been held, covering subjects from Intelligent Control Systems, Optimization, Robotics, Automation, Signal Processing, Sensors, Systems Modeling and Control and Industrial Informatics. Informatics applications are pervasive in many areas of Control, Automation and Robotics as is clearly illustrated in this conference.

The papers selected for inclusion in this book contribute to the understanding of relevant trends of current research on Informatics in Control, Automation and Robotics and cover, among others, novel contributions on robotics and aerial vehicles, output feedback control, as well as optimization control techniques for various applications.

We like to thank all the authors for their contributions and also the reviewers who have helped ensuring the quality of this publication.

July 2021

Oleg Gusikhin
Henk Nijmeijer
Kurosh Madani

Organization

Conference Chair

Kurosh Madani   University of Paris-EST Créteil (UPEC), France

Program Co-chairs

Oleg Gusikhin   Ford Motor Company, USA
Henk Nijmeijer   Eindhoven University of Technology,
                              The Netherlands

Program Committee

Eugenio Aguirre	University of Granada, Spain
Rudy Agustriyanto	University of Surabaya, Indonesia
Carlos Aldana	University of Guadalajara, Mexico
Manuel Aleixandre	Tokyo Institute of Technology, Japan
Joaquin Alvarez	Center Scientific Research Higher Education Ensenada Cicese, Mexico
Mihail Antchev	Technical University - Sofia, Bulgaria
Rui Araujo	University of Coimbra, Portugal
Filippo Arrichiello	University of Cassino and Southern Lazio, Italy
Augusto Luis Ballardini	Università degli Studi di Milano Bicocca, Italy
Ramiro Barbosa	ISEP/IPP - School of Engineering, Polytechnic Institute of Porto, Portugal
Juri Belikov	Tallinn University of Technology, Estonia
Karsten Berns	University of Kaiserslautern, Germany
Mauro Birattari	Université Libre de Bruxelles, Belgium
Jean-Louis Boimond	ISTIA - LARIS, France
Magnus Boman	The Royal Institute of Technology, Sweden
Marvin Bugeja	University of Malta, Malta
Kenneth Camilleri	University of Malta, Malta
Giuseppe Carbone	University of Calabria, Italy
Enrique Carrera	Army Polytechnic School Ecuador, Ecuador
Alessandro Casavola	University of Calabria, Italy
Marco Castellani	University of Birmingham, UK

Qiangda Yang	Northeastern University, China
Qinmin Yang	Zhejiang University, China
Wenxian Yang	Newcastle University, UK
Jie Zhang	Newcastle University, UK
Cezary Zielinski	Warsaw University of Technology, Poland
Primo Zingaretti	Università Politecnica delle Marche, Italy

Invited Speakers

Ilya Kolmanovsky	University of Michigan, USA
Feng Lin	Wayne State University, USA
Nuno Lau	Universidade de Aveiro, Portugal
Mireille E. Broucke	University of Toronto, Canada

Additional Reviewers

Tamas Becsi	BME, Hungary
Larissa Driemeier	University of Sao Paulo, Brazil
Junkai He	IRT System X, France
Krzysztof Łakomy	Poznan University of Technology, Poland
Marcin Nowicki	Poznan University of Technology, Poland
Radoslaw Patelski	Poznan University of Technology, Poland
Zoltan Szabo	SZTAKI, Hungary
Biyun Xie	University of Kentucky, USA

Contents

Intelligent Control Systems
and Optimization

Approximation Methods and Reference Values for Maximum Allowed Collaborative Operating Speeds in Quasi-Static and Transient Contact Cases

Christopher Schneider[1]([✉]) [iD], Thomas Suchanek[1] [iD], Martina Hutter-Mironovová[1] [iD], Mohamad Bdiwi[2] [iD], and Matthias Putz[3]

[1] Yaskawa Europe GmbH, 85391 Allershausen, Bavaria, Germany
{christopher.schneider,thomas.suchanek,
martina.hutter}@yaskawa.eu
[2] Fraunhofer IWU, 09126 Chemnitz, Saxony, Germany
mohamad.bdiwi@iwu.fraunhofer.de
[3] Fraunhofer Headquarter, 80686 Munich, Bavaria, Germany
matthias.putz@zv.fraunhofer.de

Abstract. This paper presents and utilizes suitable measurement setups for quasi-static and transient contact situations to conduct force and pressure measurements. Presented at the example of lathe machine tending, two typical risk situations are defined that represent a collision with the hand and the shoulder. Empirical measurement studies are executed with a selected cobot to model the maximum allowed collaborative speed behavior. Various influencing factors, such as sensor sensitivity, workpiece mass, robot pose, and additional padding, are assessed regarding their effect on the allowed velocity. As a result, an approximation equation is proposed to calculate allowed cobot velocities for the quasi-static case, while suitable reference values for the transient contact are given. Since a prototypical cell is required to conduct the force and pressure measurements, a priori estimation of realistic allowed operating speeds is impaired, resulting in imprecise cycle time assessment. With the presented approach, realistic values can be obtained based on a minimum of input data to refine the data accuracy at the project's beginning. This leads to higher data reliability to make sound return on investment decisions upfront.

Keywords: Biomechanical thresholds · Collaborative robots · Force and pressure measurements

1 Introduction

Fenceless machine tending states a common use case for collaborative robots (cobots), especially for small and medium-sized companies. Despite the easy-to-use character of cobots, implementation obstacles occur when it comes to the safety requirements for direct human-robot interaction in terms of the allowed forces and pressures, according to

O. Gusikhin et al. (Eds.): ICINCO 2021, LNEE 1006, pp. 3–17, 2023.
https://doi.org/10.1007/978-3-031-26474-0_1

ISO/TS 15066. To execute the mandatory measurements, a prototypical cell must be set up, including the robot paths to identify potential risk areas. Since the occurring forces and pressures mainly characterize the allowed collaborative operating speed and determine therefore the cycle time, preliminary economic considerations of the work cell are impaired. Furthermore, system comparisons to fenceless industrial robots with external safety scanners are not feasible due to lacking cycle time approximation tools. End-users and system integrators lack reference values of feasible collaborative operating speeds to assess the safety and economic-based suitability of a workpiece for collaborative applications.

This paper contributes to the system behavior analysis of force and pressure distributions in collaborative applications. Besides the transient contact with the shoulder, presented in [17], the quasi-static case (clamping contact) with the back of the hand (non-dominant side) is emphasized in this paper. As an application example, lathe machine tending is used due to the manageable spectrum of rotary-symmetric workpieces. The cobot model Yaskawa HC10 DT IP67 has been used for in-depth behavior analysis for practical suitability. The proposed test setups can be replicated to execute similar studies with other cobot models, workpiece characteristics, and applications to create a broader information database.

The presented research aims to provide a tool to approximate the maximum allowed collaborative speed in quasi-static cases based on workpiece characteristics. With an empirical study, the main influencing factors on the collaborative velocity as well as the robot's behavior can be modeled. Based on this data, a statistically valid approximation equation is presented that gives, along with the reference values for the transient contact, a sound information base to execute cycle time studies and conclude the economic feasibility of a cobot automation project. Such reference values can drastically decrease the future measurement and risk assessment efforts.

2   Theoretical Background

According to ISO/TS 15066, quasi-static and transient contact situations are differentiated [1]. While quasi-static ones describe a clamping situation in which no recoil of the affected body region is possible, transient contacts can be defined as collisions in free space with the possibility of recoil. To examine the risk associated with such a contact, empirical human subject studies revealed threshold values for pain entrance levels for allowed forces and pressures of both contact types. These values are defined for various body regions and serve as a reference for comparison with the measured values for the individual use case. It must be noted that collisions with the body regions "skull and forehead" and "face" must be avoided due to their criticality. As a countermeasure, organizational instructions or space limitations should be used in the risk assessment. The used velocity must be adjusted iteratively to match the prescribed force and pressure thresholds. Further studies for refining and expanding these values were presented in [2]. For the described measurements, the complete system consisting of robot, gripper, and workpiece is analyzed in terms of risk potentials. Suitable designated devices are available from different manufacturers.

While the force is measured over time using an integrated load cell, pressure-sensitive foils indicate the pressure distribution that can be scanned and analyzed with specific software. To simulate the body region-specific properties, a combination of damping material and spring is used.

Since a prototypical cell is required to perform these tests, a priori velocity determination is not feasible. Furthermore, generalized statements about expected forces and pressures are impaired due to the individuality of each use case. Therefore, the required flexibility of collaborative robotics is difficult to achieve from a safety perspective because the measurement is only valid for a specific situation and workpiece. Adaption to other workpieces in dynamic environments, as required in the high mix – low volume production and described by [3], would require continuous reassessment (Fig. 1).

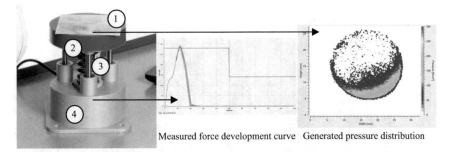

Measured force development curve   Generated pressure distribution

Fig. 1. Force and pressure measurement device according to ISO/TS 15066: 1 – Pressure indicating film, 2 – Damping material, 3 – Spring, 4 – Load cell [4].

For collision characterization, a multi-phase procedure with different classification criteria has been presented in [5]. In [6], the power flux density serves as a metric by incorporating the energy transfer and contact duration. [7] presents a three-dimensional collision force map relative to the operating space of the robot based on empirical measurements with two different cobot models. Crash tests for clamping situations with industrial robots were performed by [8], focusing on the robot mass, velocity, and singularity forces. Further collision experiments were conducted by [9], with a strong emphasis on the effect of a robotic airbag on force development. To calculate the forces in quasi-static cases, [10] and [11] published respective models. Simulative approaches and virtual sensor utilization has been presented by [12] and [13].

3   Materials and Method

3.1   Introduction

This paper presents two different empirical studies, one for the quasi-static and one for the transient contact case. The empirical study for the quasi-static contact has received funding from the European Union's Horizon 2020 research and innovation program under grant agreement No 779966 and was published in [4, 14–16]. The transient contact study was published in [17].

3.2 Quasi-Static Case

Various potential risk situations have been identified within a preliminary generalized motion study of lathe machine tending use cases. The first case describes a clamping situation, also known as quasi-static contact. The start position of the motion is illustrated in transparent blue, while the end position is represented untransparent. Such a clamping situation can occur when the robot places the workpiece on the material deposit and the operator puts his hand in between the robot and provision plate. According to ISO/TS 15066, the body region "back of the hand, non-dominant side ND" is touched. Since this is a representative risk situation for many applications, such as machine tending, material handling, and pick&place, the presented results are transferrable to a broad range of industrial use cases.

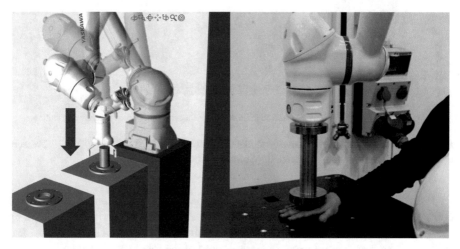

Fig. 2. Quasi-static contact case [4].

It must be noted that Fig. 2 illustrates the dominant machine tending solution with a dual gripper, while this research emphasizes single grippers to limit the number of possible influencing factors. This limitation applies to all presented contact cases and needs to be extended in future research to include the influence of the gripper mounting angle on the collaborative velocity.

Figure 3 shows a custom measurement setup that has been specifically designed for this type of contact case. To minimize natural oscillation of the system, the robot pedestal, including the cobot and a rigid frame, have been installed mutually on a solid metal plate. An adapter plate installed on the top plate of the frame serves as mounting for the measurement setup, while different drilling patterns in the plate enable flexible fixing of the measurement device on three positions. The programmed robot path is a linear motion, reaching from a position above the measurement device downwards to a point below the measurement device. When the cobot collides with the device, force and pressure are measured.

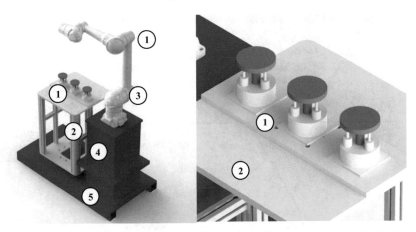

Fig. 3. Setup for quasi-static measurements: 1 – Adapter plate with measurement device, 2 – Rigid frame, 3 – Collaborative robot, 4 – Robot pedestal, 5 – Base [4].

The following Ishikawa diagram gives an overview of the collected potential influencing factors on the allowed collaborative operation speed. Clustered by the 5 M's machine, method, material, (hu)man, and measurement, the system complexity can be comprehended visually. Based on the presented Ishikawa diagram, the criteria have been specified with respective characteristics, summarized in Table 1. The pose and sensor sensitivity of the robot were considered machine-dependent factors. Individualities in the kinematic chain and the robot behavior close to singularity positions influence the sensor behavior. Therefore, measurements at different poses were conducted using the percental reach utilization of the respective measurement point as a metric. For this research, a close position at 40% of the maximum reach, a middle position at 56% and a stretched position at 72% was used. For the used robot model, an effective reach (distance between base and p-point) of 1200 mm is stated, leading to positions at 480 mm (close), 672 mm (middle), and 864 mm (stretched). Since this model uses torque sensors in each joint as PFL technology, the force limits were adjusted to cover different sensitivities. It must be noted that the force limit is indirectly proportional to the sensitivity extent. For the experiments, force limits of 140 N (low), 130 N (low), 100 N (middle), and 50 N (high) were used. By focusing on the body region "backside of the hand, non-dominant side" with a quasi-static force threshold value of 140 N, a matching start force limit (140 N) has been chosen accordingly. Further limits have been added to assess this force area in greater detail and cover the influence of higher sensitivities. For comparability of similar studies, the software version YAS4.12.01A(EN/DE)-00 should be used. The methodological and material-related properties of the quasi-static tests are characterized by a frontal contact with different workpieces that vary in diameter, length, material, and weight. The used combination of damping material, spring, and thickness of the layer material simulate the focused body region (Fig. 4).

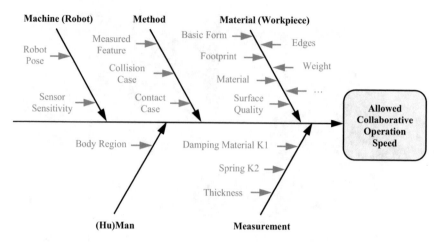

Fig. 4. Ishikawa diagram for the quasi-static contact case [4].

Table 1. Influencing factors on the allowed collaborative operating speed [4].

General criteria	Criteria	Characteristics
Robot pose	Close, middle, stretched	40%, 56%, 72%
Sensor sensitivity	Low, middle, high	140 N, 130 N, 100 N, 50 N
Software version	–	YAS4.12.01A(EN/DE)-00
Measured feature	Individual	–
Collision case	Quasi-static	–
Contact case	Frontal	
Body region	Hand and fingers non-dominant side	190 N/cm^2 pressure, 140 N force
Damping material	K1	70
Spring	K2	75 N/mm
Thickness	–	7 mm
Measured feature	Diameter	[mm]
Measured feature	Length	[mm]
Measured feature	Material	Steel, Aluminum, POM, PLA
Measured feature	Weight	[kg]

For realistic results, the following measurement conditions were used. To minimize system vibrations, the robot-pedestal assembly and the measurement frame must be mounted on a mutual massive base plate. On top of the frame, the measurement device is installed firmly on a plate to guarantee stability. To guarantee a sufficient acceleration ramp and to reach the set operating speed, the distance between the start and endpoint of the robot path must be high enough. Since this set velocity must be present during the collision, the programmed endpoint must lie below or behind the contact point at a suitable distance. Otherwise, the robot would decelerate and therefore collide at reduced speed, which distorts the measurement result. To identify the maximum allowed collaborative speed (MACS), the operating velocity is adjusted in 1 mm/s to guarantee high-resolution results. For optimal results, tool data must be configured correctly, and the torque sensors should be calibrated regularly. Since a collaborative operation is measured, the required safety functions must be activated, such as anti-clamping, retract or pushback. Furthermore, the environmental conditions should match practical standards, in this case, 21 °C and 60% humidity. The pressure-sensitive foils are specified for temperatures between 17 °C and 38 °C and humidity between 35% and 80%. To counterbalance measurement uncertainties, ten force measurements are executed for each run by excluding the maximum and minimum outlier values and averaging the remaining eight values. The resulting average value is taken as a representative value to compare with the threshold values defined in ISO/TS 15066. To guarantee optimal development of the pressure foils, 30 min between collision and scanning is recommended.

3.3 Transient Contact Case with Cobot Outer Geometry

From the previously mentioned motion sequence, a contact between the robot outer geometry and the upper body has been identified as a second risk case. During feeding and removing the workpieces from the lathe machine and when the robot moves between the machine's inner area and the door, a contact between the robot itself and the operator can occur. Figure 5 illustrates the potential contact outer geometries of the robot. By mounting the robot on a pedestal at a reasonable height of 900 mm, the motions will likely be executed at shoulder height, which is the emphasized body region for the respective experiments.

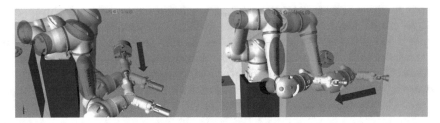

Fig. 5. Potential transient risk cases [17].

For this experiment series, the previously mentioned robot is installed on a 900 mm high pedestal bolted to the ground. A special design guarantees a realistic representation of the transient contact case by enabling free oscillation and adjustable weight to match different body region masses, in this case the shoulder (m = 40 kg). Different weight plates were installed with a screw nut on a large locating bolt with a thread. On top of the plates, the measurement device was installed using screw clamps. To minimize friction and to provide a sufficient pendulum length, this assembly was connected to the steel tracks of a 0.5 t crane. The crane tracks were adjusted to a height in which the measurement device matches a realistic height of 1450 mm. To guarantee a controlled recoil movement at a certain rebound angle, ropes were attached to the device (Fig. 6).

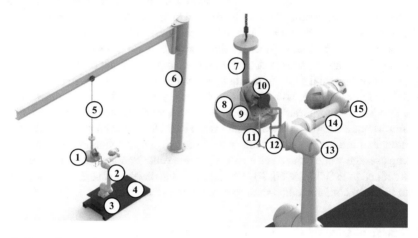

Fig. 6. Setup for transient measurements: 1 – Measurement setup, 2 – Robot, 3 – Pedestal, 4 – Base plate, 5 – Chain with hook, 6 – Crane, 7 – Locating bolt, 8 – Weight plates, 9 – Holding design, 10 – Measurement device, 11 – Screw clamp, 12 – Elbow big cap, 13 – Elbow small cap, 14 – Forearm and 15 – Wrist cap [18].

To further classify and characterize influencing factors on the transient MACS, the same procedure for the quasi-static case has been used by creating an Ishikawa diagram and assigning criteria and characteristics, which can be seen in detail in [17]. As potential contact areas on the robot side, four contours were defined: the big and small elbow caps, the forearm, and the wrist cap. To cover the worst-case scenarios that must be considered in a risk assessment, the least favorable edge geometries were used. As countermeasures, different protective measures on the collision surfaces were used by utilizing 5 mm thick neoprene and 140 mm foam padding. By this, the influence of additional padding on the transient MACS is assessed. To match the shoulder body region properties, a spring constant of $k = 35$ N/mm and 14 mm thick silicone damping material with A30 hardness was used. To utilize the maximum payload of the robot, a steel shaft with 110 mm diameter in 230 mm length was manufactured and attached to the robot flange. Iterative measurements were conducted with 100 N and 50 N force limit and a resolution of 10 mm/s for this experiment series. Further experiment specifications were similar to the quasi-static case.

4   Results

4.1   Quasi-Static Contact Case

4.1.1   Test on Gripper and Material

In the first measurement series, different chuck workpieces with 110 mm in diameter and 30 mm or 50 mm length were used that were cut from bar material or were 3D printed.

Used materials were steel, aluminum, PLA (polylactic acid), and POM (polyoxymethylene). The single workpieces were handled with a Schunk Co-act EGP-C gripper with custom aluminum jaws to hold the 110 mm workpieces with positive locking. Since this gripper is designed for small workpieces, the 50 mm long steel workpieces were not held reliably. Therefore, a screw clamp was attached that increased the gripping force, leading to asymmetric mass distribution. For that reason and for the sake of general applicability, a representative solution was required that simplifies the setup and makes it gripper model-independent. As a result, an aluminum reference workpiece was manufactured that matches the mass and length properties of the flange-gripper-workpiece assembly (Fig. 7).

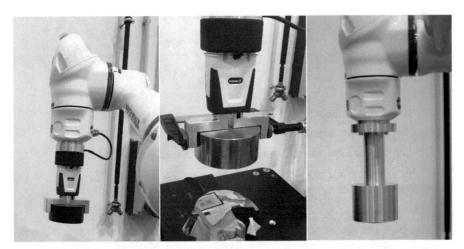

Fig. 7. Selected test setups for gripper and workpiece: Left – Schunk gripper with 50 mm long POM material, Middle – Schunk gripper with 50 mm long workpiece and screw clamp, Right – Aluminum basis workpiece [4].

Table 2 shows the mass difference of 158 g between the combination "Schunk gripper and 30 mm long workpiece" and the aluminum reference workpiece (highlighted in green), making both setups comparable in terms of mass. A similar reference workpiece was designed and manufactured with steel that covers the maximum possible load of

Table 2. Weights of the tested workpieces for the D110 mm measurement series [4].

Material	PLA		POM		Aluminum			Steel		
D [mm]	110		110		110			110		
L [mm]	50	30	50	30	50	30	BW	50	30	BW
m_1 [kg]	0.148	0.097	0.674	0.403	1.34	0.8	3.584	4.449	2.222	6.796
m_2 [kg]	1.352	1.301	1.878	1.607	2.544	2.004		5.653	3.426	

the robot. As a result, the following workpieces and masses were used for the first measurement series.

To evaluate the comparability between the gripper-workpiece combination and the reference workpiece, MACS values for the 110 mm diameter for the close, middle, and stretched positions and 140 N, 130 N, 100 N, and 50 N were determined (Fig. 8).

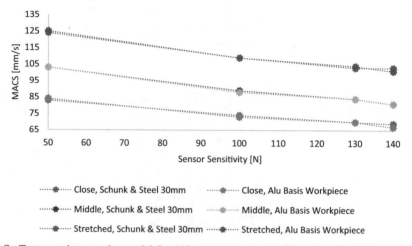

Fig. 8. Test on gripper and material for 110 mm: comparison of Schunk Gripper with 30 mm steel workpiece and aluminum basis workpiece [4].

Since both measurement series show only minimal deviation from each other, the gripper does not influence the measurement as long as reliable holding is guaranteed and the gripper-jaw-workpiece can be considered stiff. In this case, only the mass properties influence the measurement, which will be analyzed in more detail. To simplify the experiment design, the following measurements with 80 mm, 50 mm, and 20 mm diameter will be executed with the aluminum and steel workpiece as well as a combination of 50 mm long POM material with a Schunk gripper. For comparable results, the workpieces and the reference workpiece are turned on a lathe machine by some millimeters to the target diameter. Consequently, the mass properties of the reference workpieces with 110 mm, 80 mm, 50 mm, and 20 mm vary only slightly in mass since the main body remains. Table 2 summarizes the used masses (Table 3).

Table 3. Weights of the tested workpieces for the D80 mm to D20 mm measurements [4].

Material	POM		Aluminum BW		Steel BW		
D [mm]	80	50	80	50	80	50	20
L [mm]	50	50					
m_1 [kg]	0.54	0.423	–	–	–	–	–
m_2 [kg]	1.744	1.627	3.484	3.414	6.483	6.195	6.041

4.1.2   Test on Sensor Sensitivity, Robot Pose, and Diameter

The sensor sensitivity's influence on the MA was measured for all workpiece diameters from 110 mm to 20 mm. As can be seen in Fig. 9, the robot retracts already after about 180 ms within the transient contact area. Therefore, an actual clamping does not occur, and higher threshold values are acceptable under the utilization of the multipliers defined in ISO/TS 15066.

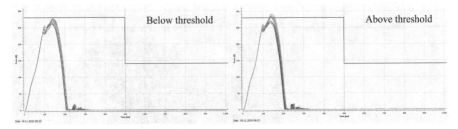

Fig. 9. Force development behavior for quasi-static contact cases [15].

The measurement data shows that the allowed velocity increases with higher sensor sensitivity (decreasing force limit in N) with nearly linear progression. The data set is separated into three data bundles, representing one position each. Even when comparing the overall minimum and maximum MACS values over diameter, the diameter does not show a specific pattern. Consequently, the diameter can be excluded as a potential influencing factor from a force perspective within the area of 110 mm to 20 mm. The pressure perspective is focused on in the following subchapter.

4.1.3   Test on Tool and Workpiece Weight

By analyzing the measurements with different workpieces from 110 mm to 20 mm, an increasing weight impact can be seen, especially depending on the robot pose and sensor sensitivity. Due to the low lever and its effect on the kinematics and sensor behavior, the close position is nearly unaffected by increasing weight, while this factor takes greater effect with an increasing lever on the middle and stretched position by decreasing the MACS. Higher sensor sensitivity (lower force limit in N) compensates for the weight impact better: while the MACS at the close and middle position is nearly equally compensated, the MACS is highly negatively affected at the stretched position. By analyzing the minimum and maximum MACS value deviation by robot pose for the different sensitivity settings, it can be seen that the highest sensitivity (lowest force limit) of 50 N compensates best independent of the robot pose. For lower sensitivities, this deviation increases exponentially with increasing range.

4.1.4   Test on Pressure

To emphasize smaller diameters with a worst-case pressure distribution, experiments with 50 mm and 20 mm diameter steel and aluminum workpieces were executed using pressure foils with a suitable resolution. As velocity test values, the force-based MACS values were used, while the pressure results serve as validation, indicating either a compliant speed or the need to reduce the velocity. As expected, the 110 mm diameter did not show any pressure results since the contact surface matches the measurement device diameter. 80 mm diameter did not show significant results either due to an even pressure distribution. First results were obtained at 50 mm diameter with compliant pressure values. At 20 mm, a first exceeding of the pressure threshold values was observed. Exemplary pressure distributions of 50 mm and 20 mm are illustrated in Fig. 10.

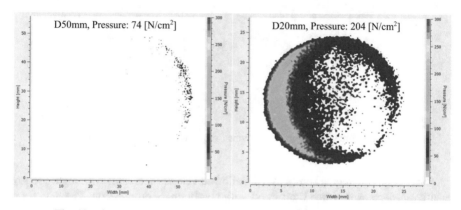

Fig. 10. Exemplary pressure foil results for 50 mm and 20 mm diameter [4].

Since the foils were inserted differently in the scanner, both pictures vary in the arrangement. For the 20 mm measurement, an unequal distribution can be seen, resulting from unideal experiment conditions. Even though the robot flange was oriented parallel to the measurement device, slight positioning inaccuracies led to a higher pressure application on one side. Such conditions also occur in practice when measuring industrial use cases and represent, therefore, an accurate picture of practical procedures. To identify the influence of the workpiece's tilting angle on the pressure distribution result and its compliance with the defined threshold values, the robot program was adjusted by modifying the rotation around the respective axis in 0.5 steps. At +0.5, the lowest pressure value was identified that complies with the threshold and therefore validates the force-based MACS values. Tilting at −1.5, −0.5, +1 and +1.5 led to non-compliant values. Therefore, it can be stated that the identified force-based MACS values were validated by the pressure measurements at optimal conditions, leading to the exclusion of the pressure as significant influencing factors for diameters from 110 mm to 20 mm, since the force is the dominant criterion. For diameters below 20 mm, however, only pressure must be considered.

4.1.5   Concluded MACS Approximation Equation

To facilitate the MACS identification for quasi-static cases, an approximation equation has been derived by statistical analysis of all 222 measurement series. As a database, the MACS values with assigned values for position, force limit, and attached mass were used. Multiple linear regression revealed the following approximation equation with the parameters position p [%], sensor sensitivity s [N] and workpiece mass m [kg] and a coefficient of determination of $R^2 = 0.964902$:

$$MACS = 53.1107 + 1.15885p - 0.2283s - 0.4578m \qquad (1)$$

The overall context is illustrated in Fig. 11 as a three-dimensional diagram including the MACS, force limit, position, and reach.

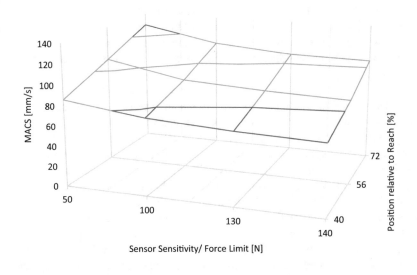

□ 0-20   □ 20-40   □ 40-60   □ 60-80   □ 80-100   □ 100-120   □ 120-140

Fig. 11. Maximum allowed collaborative speed relative to sensor sensitivity setting and robot pose for workpiece diameters 110 mm to 20 mm [4].

4.2 Transient Contact Case with Cobot Outer Geometry

For the transient contact case, four experiments were performed with the outer geometries elbow big cap, elbow small cap, forearm, and wrist cap as well as without and with the protective measures neoprene and foam. Overall, the padding showed a significant effect only for one contour, while having a small effect on the others. This can be explained by the rounded design of the cobot according to ISO/TS 15066. The force limit of 50 N did show higher transient MACS values compared to the more insensitive 100 N sensor setting. Only for one experiment, 50 N was not feasible due to a self-triggering of the sensors. Consequently, it can be stated that additional padding of edges can benefit the maximum allowed collaborative speed if critical areas are present by distributing the pressure equally. As a result, transient MACS values between around 700 mm/s up to around 900 mm/s can be achieved in collaborative mode with optimal outer contours, making cobots a reasonable solution in terms of cycle time in fenceless production.

5 Conclusions and Discussion

This paper contributes to fundamental research to understand the main influencing factors on the force and pressure distribution in collaborative applications for quasi-static and transient contact cases. As potential contact body regions, the back of the hand, non-dominant side, and the shoulder were used. A broad empirical study of quasi-static contacts with a selected cobot model revealed the following main influencing factors on the maximum allowed collaborative speed: robot pose, sensor sensitivity, and workpiece weight. Based on this database, an approximation equation has been derived using statistical analysis. For the transient contact, different outer contours of the cobot were measured to determine reference values for the maximum allowed collaborative speed. Furthermore, the effect of additional padding for an improved pressure distribution has been emphasized.

For this research, single grippers were assumed for simplification reasons, while double grippers are the dominant solution in machine tending. In future research, the influence of double grippers and the respective installation angle must be identified. Since the used cobot model utilizes torque sensors as safety technology, other technology options such as motor current monitoring or sensitive skins must be assessed in future research. For the transient case, measurement inaccuracies occurred due to the oscillation of the measurement design, which should be counterbalanced in the following research.

With the proposed test setups for both cases, multiple factors can be identified and analyzed with empirical studies. To supplement the presented database, other cobot models and workpiece examples can be used in future studies. The proposed equation and reference values serve as a database for preliminary cobot velocity and cycle time assessments. Consequently, efforts for risk assessment and certification can be drastically reduced for improved efficiency in cobot cell planning. Higher data reliability supports the planner in estimating cycle times and therefore evaluating the economic feasibility of a potential automation project upfront.

References

1. ISO/TS 15066: Robots and robotic devices - Collaborative robots. International Organization for Standardization (2017)
2. Behrens, R., Pliske, G.: Fraunhofer IFF, Otto von Guericke University Trauma Surgery Clinic: Human-Robot Collaboration: Partial Supplementary Examination (of Pain Thresholds) for Their Suitability for Inclusion in Publications of the DGUV and Standardization, 4 July 2021
3. Eder, K., Harper, C., Leonards, Z.: Towards the safety of human-in-the-loop robotics: challenges and opportunities for safety assurance of robotic co-workers. In: The 23rd IEEE International Symposium on Robot and Human Interactive Communication, Edinburgh, UK, 25–29 August 2014, pp. 660–665 (2014)
4. Schneider, C., Rahman, S., Seizmeir, M., Suchanek, T., Klos, M., Hutter-Mironovová, M.: COVR: COVR Project RACOS: Risk Assessment Metamodel for Cobot Operating Speed Determination - Milestone 2, 27 July 2021. https://covrfilestorage.blob.core.windows.net/documents/casestories/COVR%20RACOS%20-%20Milestone%202%20Report.pdf
5. Haddadin, S., De Luca, A., Albu-Schäffer, A.: Robot collisions: a survey on detection, isolation, and identification. IEEE Trans. Rob. **33**(6), 1292–1312 (2017)

6. Vemula, B., Matthias, B., Ahmad, A.: A design metric for safety assessment of industrial robot design suitable for power-and force-limited collaborative operation. Int. J. Intell. Robot. Appl. **2**(2), 226–234 (2018). https://doi.org/10.1007/s41315-018-0055-9

7. Svarny, P., Rozlivek, J., Rustler, L., Hoffmann, M.: 3D collision-force-map for safe human-robot collaboration (Preprint) (2020)

8. Haddadin, S., Albu-Schäffer, A., Hirzinger, G.: Safe physical human-robot interaction: measurements, analysis and new insights. In: Kaneko, M., Nakamura, Y. (eds.) Robotics Research, pp. 395–407. Springer, Cham (2010). https://doi.org/10.1007/978-3-642-14743-2_33

9. Weitschat, R.: Industrial human-robot collaboration: maximizing performance while maintaining safety. Doctoral thesis, Universität Rostock, Lehrstuhl für Mechatronik, Fakultät für Maschinenbau und Schiffstechnik, Rostock, Germany (2019)

10. Ganglbauer, M., Ikeda, M., Plasch, M., Pichler, A.: Human in the loop online estimation of robotic speed limits for safe human robot collaboration. Procedia Manuf. **51**, 88–94 (2020)

11. Kovincic, N., et al.: A model-based strategy for safety assessment of a robot arm interacting with humans. In: Proceedings in Applied Mathematics & Mechanics (PAMM), vol. 19, no. 1 (2019)

12. Shin, H., Kim, S., Seo, K., Rhim, S.: A real-time human-robot collision safety evaluation method for collaborative robot. In: 3rd IEEE International Conference on Robotic Computing (IRC), Naples, Italy, 25–27 February 2019, pp. 509–513 (2019)

13. Yen, S.-H., Tang, P.-C., Lin, Y.C., Lin, C.-Y.: Development of a virtual force sensor for a low-cost collaborative robot and applications to safety control. Sensors **19**(11), 2603 (2019)

14. Schneider, C.: COVR: COVR-Case story: Creating a Metamodel for Determining Safe Cobot Speeds, 27 July 2021. https://covrfilestorage.blob.core.windows.net/documents/casestories/COVR%20Case%20Story%20-%20RACOS%20Yaskawa%20v4.pdf

15. Schneider, C.: COVR: COVR-Appendix, 27 July 2021. https://covrfilestorage.blob.core.windows.net/documents/casestories/COVR%20RACOS%20-%20Milestone%202%20Report%20-%20Appendix_compressed.pdf

16. Schneider, C., Suchanek, T., Klos, M., Hutter-Mironovová, M.: COVR: COVR Project RACOS: Risk Assessment Metamodel for Cobot Operating Speed Determination-Milestone 1, 27 July 2021. https://covrfilestorage.blob.core.windows.net/documents/casestories%5CRACOS%20Milestone%201.pdf

17. Schneider, C., Seizmeir, M., Suchanek, T., Hutter-Mironovová, M., Bdiwi, M., Putz, M.: Empirical analysis of the impact of additional padding on the collaborative robot velocity behavior in transient contact cases. In: Proceedings of the 18th International Conference on Informatics in Control, Automation and Robotics-ICINCO, pp. 216–223 (2021). ISBN 978-989-758-522-7; ISSN 2184-2809. https://doi.org/10.5220/0010604202160223

18. Schneider, C.: Robotic automation of turning machines in fenceless production: a planning toolset for economic-based selection optimization between collaborative and classical industrial robots. Doctoral Thesis, Chemnitz University of Technology, Professorship Production Systems and Processes, Chemnitz (2022)

Toward Real-Time Multi-objective Optimization for Bus Service KPIs

Nabil Morri[1,3]([✉]) [iD], Sameh Hadouaj[2,3] [iD], and Lamjed Ben Said[3] [iD]

[1] IT Department, Liwa College of Technology, Abu Dhabi, UAE
morrynabil@gmail.com
[2] Computer Information Systems Department, Higher Colleges of Technology, Abu Dhabi, UAE
[3] SMART Lab, Institut Supérieur de Gestion de Tunis, Université de Tunis, Tunis, Tunisia
lamjed.bensaid@isg.rnu.tn

Abstract. Cities are currently facing many transport problems and deploying intelligent transport systems that play an important role in urban traffic management, especially public transport. With the evolution of public transport systems and the advanced technologies they use, there is a great transformation in public transport control systems. Transit system data is also highly dynamic and voluminous, hence the great need for intelligent control systems to better exploit this data and provide better performance. The contribution of this work is to model and implement an intelligent public transport system to detect and manage disturbances while respecting the constraints of the traffic situation. This system combines various measures of Key Performance Indicators (KPIs) into a single performance value, covering several dimensions depending on the type of service quality to be guaranteed. Particular attention has been focused on the method used in the formulation of the multi-objecvtive optimization model and the determination of the weighting factors. Multi-agent modelling is adopted in the design of our system. The experiments represent real disturbances observed in public transport networks. The results explain how the resolution strategy of our model outperforms that of the real world.

Keywords: Multi-agent simulation · Multi-objective optimization · Intelligence control system · Public transportation · Key performance indicator

1 Introduction

In the economic, environmental, and political context, public transport has many advantages. It saves on the fuel costs of individual cars financially and ecologically, reduces congestion on roads and in parking areas, and is also safer than so-called private transport. Public transport offers a wide variety of resources. There are different modes of transport with an infrastructure of multiple trip lines, stations, roads, etc. These resources will have to be well exploited to ensure good quality of service and especially against disturbances. With the evolution of today's public transport systems and advanced technologies, there is a great transformation in public transport control systems. Indeed, transport managers are using intelligent systems such as automatic vehicle location systems and new communication technologies to better manage disturbances.

O. Gusikhin et al. (Eds.): ICINCO 2021, LNEE 1006, pp. 18–36, 2023.
https://doi.org/10.1007/978-3-031-26474-0_2

In the public transport system, a disturbance is an event that occurs suddenly and changes the traffic status of the traffic network into a situation that is generally unsatisfactory in quality of service. The complexity of the traffic network means that several disturbances can occur at the same time, or that one disturbance can lead to others. In addition, the data of public transport systems are very dynamic and voluminous, hence the great need for intelligent systems to better exploit the resources of transport systems and to provide the best performance.

Consequently, public transport control systems must quickly detect disturbances and adapt to new situations in order to improve the quality of service through an optimization of performance measures. These measures represent Key Performance Indicators (KPIs). The transit system must provide comparative information that allows the control system to identify performance gaps and set measures and targets to resolve them. Extending [1], fuzzy approach has been applied to determine the weights of used KPIs in our Multi-Objective linear Programming (MOLP) problem. In this approach each indicator is weighted according to its contribution that is expressed as a function of its membership function. This paper develops the fuzzy solution approach using linear membership functions for each indicator and shows how it is benefits on global transit performance and waiting time of the passengers.

Furthermore, multi-agent modelling provides a solution fitted to the activities of public transport networks where autonomous entities (buses, stations, itineraries, etc.), called agents, dispersed in a dynamic environment which is the traffic of the transport network. They adapt their behaviors to the perturbation they perceive and interact with each other to perform optimal control action.

In [2] the authors present an overview of agent-based models and its applications on the field of the transportation planning, and discuss the challenges needed to be overcome for the further growth of agent-based modelling in the field of transport.

In addition, more recently, an extensive study has been presented by the authors of [3] on several traffic and transportation agent-based applications. According to this study, some previous attempts at multi-agent applications in the urban transit sector date back to the 1990s. For example, we cite the work of [4] that proposed a cooperative agent-based architecture as a solution to improve the management and the control of transportation traffic.

Without surprise, it was also a time when a lot of research was underway to define the real field of application of multi-agent systems. Much research from different fields of computer science and even artificial intelligence were trying to determine if agents were different to objects, from the design point of view, or to autonomous processes, from the programming point of view. While members of the AI community initially took advantage of the complex and dynamic nature of transportation systems to develop and support agent theory, transportation engineers and practitioners have now begun to recognize the natural ability of the multi-agent metaphor to model transportation traffic phenomena. Because of their characteristics and concepts, multi-agent systems have a natural ability to deal with a wide range of problems in traffic management including control system in public transportation [5].

Most multi-agent work on transportation traffic management makes transit systems more autonomous and responsive to perturbations, such as the work of [6]. The authors describe and compare integrated TRYS and TRYS autonomous agents, two multiagent systems that perform decision support for real-time traffic management.

The modelling of intelligent traffic systems has also been studied in [7–9]. These studies focus on similar applications, namely the regulation of freight transportation and the optimization of resource.

Another study has been reported in the literature, which provides a fairly good survey of the application of multi-agent approaches to transportation logistics [10]. Nevertheless, the challenge of modelling more realistic control system in transportation traffic has encouraged the increasing use of intelligent agents. For example, drivers are endowed with cognitive abilities to detect traffic congestion and plan a trip that represents a mental model of the environment and an expectation of the utility of their choices (e.g. [11, 12]).

In the same sense, agent concepts have also proven to be very useful in the analysis of travel demand and advanced passenger information systems [13].

Through this literature, we observe that the agent concept has been very influential in the development of the modelling of intelligent transportation systems, including transit control systems. Hence, Multi-agent modelling can provide a solution adapted to the activities of public transport where autonomous entities, called agents, interact with each other in an environment: (i) distributed: information is geographically dispersed on the network requires distributed agents, (ii) open: manage entities that can freely enter and exit the environment such as vehicles can enter and exit the road network, (iii) dynamic: there is a daily change of information, e.g., appearance of perturbation. (iii) heterogeneous: there are various types of actors such as vehicles, stations, links between stations, etc. Moreover, these entities can reason, communicate via messages to resolve conflicts, and reach the best solution. These characteristics [14] prove that the use of the multi-agent approach in the modelling of our control system introduces a more flexible and efficient representation in the processes it models. We describe on details our multi-agent modelling of our control system.

Our objective is to model and implement a multi-agent control system that manages perturbations. It detects and finds the best control action while respecting the traffic condition. This system combines various KPI measures into a single performance value, covering several dimensions depending on the type of service quality.

In the following, Sect. 2 presents a literature review of KPIs. Section 3 describes how to deal with disturbances with our proposed control system. Section 4 details the KPIs formulas. Section 5 defines the multi-objectives optimization resolution and the proposed fuzzy logic approach. The Sect. 6 presents the multi-agent model for the system by describing the agents with their exchange of messages. Section 7 shows some computational results for the problems, which are obtained through a simulation on a real network in the city of Portland in Oregon. In Sect. 8, we conclude and provide some perspectives.

2 Literature Review of KPIs

There is an extensive literature on various aspects of KPIs. [15] evaluates performance by EWT: Excess Wait Time, AWT: Actual Wait Time and SWT: Scheduled Wait Time.

Moreover, in [16] this performance is defined by the average waiting time expected by passengers. [17] defines performance by the observed time interval deviations between trips of the same line compared to the regular frequency of vehicles during a given period.

[18] describes the Gini index as another indicator in the form of a forward regularity index. [19] describes regularity as an index on vehicle entries at stations.

Other projects define another indicator which is punctuality as a determining criterion in the final performance formulation. Punctuality is defined in [20] as a comparison of actual departure times and scheduled departure times at the station. In [21] the authors distinguish three measures of punctuality. PIR: Punctuality Index based on Routes, DIS: Deviation Index based on Stops and EIS: Evenness Index based on Stops. However, in [22], the author converts the punctuality index to a percentage to define the proportion of the trip that was punctual.

In [23], three alternative performance measures are proposed: EI: Earliness Index, WI: Width Index and SSD: Second-Order Stochastic Dominance Index. These indices are used in two forms to measure the unreliability of a bus service: (i) the distribution of lane deviations for frequent services and (ii) the distribution of delays for infrequent services.

Other work, such as of [24], add another indicator called transfer time that covers the time spent when the passenger is waiting for the vehicle while changing the line at a transfer station. [25] details and explains the Input Buffer Time (IBT) formula that can be used to understand the additional unreliability caused by an incident. The authors of [26, 27] discuss another indicator called "Dwell"" which is the parking time of vehicles. "Dwell" can also be used to hold for traffic to be restored [28, 29].

From this literature, we note that there is no standard meaning for specifying and formulating performance indicators. The challenge in defining key performance indicators is to select those that will sufficiently satisfy the overall performance of public transport.

In this work, the KPIs are selected on the basis of the indices of the "operational efficiency" objective. According to literature reviews by [30], this objective will be focused on three indicators: punctuality, regularity, and correspondence. The other indicators that are related to the costs of transport, such as oil consumption and the number of kilometers traveled, are not included in our study, because our control system is oriented towards the user of the transport system and not the operator, and they are irrelevant for the passengers.

The formulas for these measures are taken from [31] and some other research works such as [20, 32], and [24]. They represent a standard that can evaluate overall traffic performance in transportation engineering and Intelligent Transportation Systems.

3 Control Method

3.1 General Control Method

To improve service performance, it is necessary to optimize the adequacy between planned and actual operations, as shown in Fig. 1. In his study, [33] values the application of control methods in control systems as guidelines to control the variability of planned versus actual operations. In this section, three main types of system control are analyzed. Particular attention will then be focused on the control method used to improve service performance in the public transport sector.

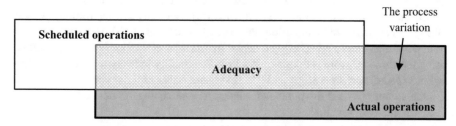

Fig. 1. Perturbation: inadequacy between scheduled and actual operations in study [1].

Control methods provide guidance on how to deal with disturbances. They detect and reduce process variability. The general science of the process control system focuses on the function and information flows of operations. For this reason, its principles are also suitable for process control theory and the description and mitigation of system variability [34]. Again, the authors of [35] stated that the ultimate goal of process control theory is to maximize performance while keeping the process practical.

[36] presents the three main control methods illustrated in Figs. 2, 3 and 4. These control methods were designed to mitigate the negative effects of disturbances on the process and achieve the desired results.

Use of a Buffer. It consists in providing resources in the form of a measurement buffer allowing to detect the perturbation without generating repercussions on the desired result.

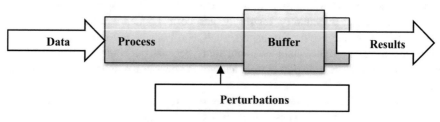

Fig. 2. Use of a Buffer method.

Feedback. In case of unacceptable variation between scheduled and actual operations, a regulator intervenes to restore the desirable situation.

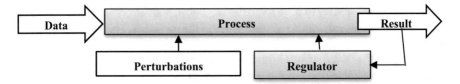

Fig. 3. Feedback method.

Feed-forward. Instead of observing the trends of the variation, the regulator prepares the regulation action based on the simulation.

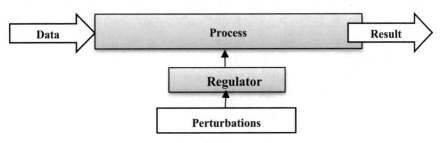

Fig. 4. Feed-forward method.

3.2 Proposed Method

The three methods have specific advantages and disadvantages. In practice, they can also be used in combination. In our study, the control method is mainly used to evaluate current performance and to adjust operations in case of inadequate performance.

Therefore, it is essential to have the resources for detecting the perturbation and a regulator to reduce the variation as much as possible. Consequently, in our model it is necessary to have a buffer for the detection. While, after detection the selection of the control action is based on the calculation of the performance measurements of several regulation maneuvers. This is the feedback method (Fig. 5).

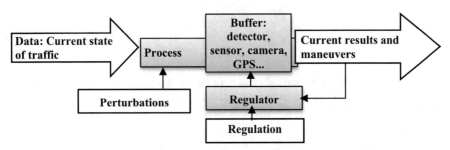

Fig. 5. Proposed Control method [1].

As well, it is necessary to have scenarios of control maneuvers in memory to be able to compare the predicted results with the results of the scenarios. This is the feed-forward method.

Therefore, the three control methods that are mentioned above are used in combination in the modelling of our control system.

4   The KPIs Formulas

In particular, operational efficiency is the ability of the transport system to meet a theoretical schedule. It is a question of maintaining the scheduled passage times of vehicles at the stations and not the travel times and speeds of the trips. For this purpose, the evaluation is done on the travelled part of the network and not on a forecast of the total travel time of the trip. The following is a selection of operational efficiency KPI measures.

4.1 Punctuality Index

Punctuality is defined in [20, 23] as a comparison of the actual arrival times with the scheduled departure times at the station. Its formula is:

$$I_{PUN} = \frac{S_1^2}{PUN^2} \text{ with } S_1^2 = \frac{1}{n} \sum_{i=1}^{n} (t_i - t_t)^2 \tag{1}$$

- n: the number of vehicles of the same line arriving at station in a defined period.
- $\overline{PUN}$: $\frac{1}{n-1} \sum_{2}^{n-1} (PUN_i - PUN_{i-1})$ the average punctuality for the n vehicles.
- $t_i = t_{arr\_i} + Dwel_i$: the actual departure time of the i-th vehicle, while $Dwel_i$ is the time spent by the vehicle in the station to board the passengers.
- $Dwel_i = t_{Mont} * N_{Mont} - t_{Desc} * N_{Desc}$, with t_{Mont} et t_{Desc} are respectively the average time spent by the passenger to get on or off the vehicle and N_{Mont} et N_{Desc} are respectively the number of passengers to be picked up and dropped off the vehicle. According to [37] this value is calculated following a Poisson distribution since the phenomenon of arrival and departure of passengers are independent.
- t_t: the scheduled departure time of the i-th vehicle.

4.2 Regularity Index

The regularity index measures the differences in time intervals observed at the control point (a priori station) between successive vehicles of the same line compared to the scheduled frequencies. The formula of the regularity index is:

$$I_{REG} = \frac{S_2^2}{h^2} \text{ with } S_2^2 = \frac{1}{n-1} \sum_{i=2}^{n} (h_i - h_t)^2 \tag{2}$$

- $\overline{h}$: $\frac{1}{n-1} \sum_{2}^{n-1} (t_i - t_{i-1})$ the average frequency for the n vehicles.
- h_i: $t_i - t_{i-1}$ (i = 2,...I), the current time interval of the i-th vehicle.
- h_t: the scheduled time interval of the i-th vehicle.

It should be noted that there is not a regularity for the first vehicle. Therefore, the sum of the deviations is divided by the number of vehicles decreased by 1 (n−1).

4.3 Correspondence Index

The correspondence is the transfer time spent by the passenger when changing buses. Its formula is the following:

$$I_{COR} = \frac{S_3^2}{\bar{c}^2} \text{ with } S_3^2 = \frac{1}{n} \sum\nolimits_{i=1}^{n} (c_i - c_t)^2 \tag{3}$$

- c_i: the actual correspondence of the i-th vehicle.
- c_t: the scheduled correspondence of the i-th vehicle.
- $\bar{c}$: $\frac{1}{n-1} \sum\nolimits_{2}^{n-1} (c_i - c_{i-1})$ the average correspondence for the n vehicles.

The actual 'c_i' or scheduled 'c_t' of the i-th vehicle is the sum of the waiting time between vehicle 'i' and the vehicles in connection at the transfer station. It is equal to:

$$C_i = \sum\nolimits_{(i=1)}^{n} \Delta_{ij} \text{ with } \Delta_{ij} = t_i - t_j + D_{ij} \tag{4}$$

- Δ_{ij}: the remaining arrival time of the vehicle in connection 'j' with the vehicle 'i'.
- t_i: the actual arrival time of vehicle 'i'.
- t_j : the actual departure time of the vehicle in connection 'j'.
- D_{ij} : the walking time between two connecting stops of the two vehicles 'i' and 'j'.

The correspondence value Ct is calculated in the same way as the actual correspondence, using the scheduled times instead of the actual times.

5 Optimization Approach

5.1 Multi-objectives Optimization Resolution

In our problem there are several objectives which are the index measures to be achieved and there is a non-negative objective function for each measure. Furthermore, the solution space is known in advance in the form of the control where the performance of the optimal solution can be calculated mathematically. In this case, constructing a Multi-Objective linear Programming (MOLP) that combines the objective functions into a single performance function with a significant weighting for each becomes a promising solution. The objectif function of our problem, which is the performance value "F", is formulated as follows:

$$F = w_{I_{PUN}} \cdot I_{PUN} + w_{I_{REG}} \cdot I_{REG} + w_{I_{COR}} \cdot I_{COR} \text{ with } w_{I_{PUN}} + w_{I_{REG}} + w_{I_{COR}} = 1 \tag{5}$$

where $w_{I_{PUN}}$, $w_{I_{REG}}$ and $w_{I_{COR}}$ are the corresponding weight of each index. The basic principle of this approach is to define a profit function that reflects the "best" solution proposed to our MOLP problem. A technique based on fuzzy logic was invented in this context called "Objective Average Variations" (OAV). This approach gives better results than those of our previous experimentations [1]. In this approach each objective function is weighted according to its contribution. The contribution is expressed as a function of its membership function. The weights are adjusted by fuzzification to obtain the desired compromise.

Such "objectif" function can be optimized by a combinatorial method. Combinatorial optimization is a subject that consists in finding an optimal object from a finite set of objects. It operates in optimization problems where the set of feasible solutions is discrete or can be reduced to discrete. The purpose is to find the element that optimizes the objectif function.

5.2   Fuzzy Logic for Determination of Weighting Factors

Our optimization problem consists of making decisions according to several objectives which may sometimes be antagonistic. Our resolution consists of two elements: the list of objectives (indices) and the list of solutions. Each disturbance situation requires a solution which is a control action.

Consider $(A_1, A_2,..., A_n)$ the solutions represented by the predefined control actions. The objectives are defined by the list of the indices I_{PUN}, I_{REG} and I_{COR}. Table 1 shows the matrix of performance indicator measures a_{ij} calculated for each control action.

Table 1. Performance indices measures.

Control action	I_{PUN}	I_{REG}	I_{COR}
A_1	a_{11}	a_{12}	a_{13}
A_2	a_{21}	a_{22}	a_{23}
...	...	...	...
A_n	a_{n1}	a_{n2}	a_{n3}

We need to find the most optimal action A_i following a "min" of the global performance function F.

$$MinF = F_i = \sum_{j=1}^{3} w_j . a_{ij} \tag{6}$$

But first we have to find the most significant combination of weights w_j of the objective function F. In this study, it is proposed to use the degree of the objective's membership functions in the performance indicator measurement matrix. A very popular method called the centroid method of membership function simplifies the calculation of weights. It allows to obtain a net solution more quickly [38]. This method is often used for real-time applications, where the computation time is important. Based to this method, we use the Eq. (7) to calculate the linear membership function of the minimization objective of F.

$$\mu(x) = \begin{cases} 1 \; if \; f_k < f_k^* \\ \frac{[f_k^* - f_k(x)]}{[f_k^* - f_k']} \; if \; f_k^* \leq f_k(x) \leq f_k' \\ 0 \; if \; f_k(x) > f_k' \end{cases} \tag{7}$$

where f_k^* and f_k' are the maximum and minimum values of the objectives. They are calculated from the ideal and anti-ideal k-th objective values. Table 2 shows a matrix containing the membership functions of the indices that represent the degrees of variation in performance.

Table 2. Matrix of membership functions of indices.

Control action	I_{PUN}	I_{REG}	I_{COR}
A_1	$\mu_1(I_{PON})$	$\mu_1(I_{REG})$	$\mu_1(I_{COR})$
A_2	$\mu_2(I_{PON})$	$\mu_2(I_{REG})$	$\mu_2(I_{COR})$
...	...	...	...
A_n	$\mu_n(I_{PON})$	$\mu_n(I_{REG})$	$\mu_n(I_{COR})$

If the $\text{Avg}_i = \frac{1}{n}\sum_{j=1}^{n} \mu_j(I_i)$ represents the average values of each performance index, then the weights w_i are calculated as follows:

$$w_{\{I_{PUN}, I_{REG}, I_{COR}\}} = \frac{\text{Avg}_{\{I_{PUN}, I_{REG}, I_{COR}\}}}{\sum \text{Avg}_{\{I_{PUN}, I_{REG}, I_{COR}\}}} \tag{8}$$

5.3 Formulation of Constraints

The following constraints are based on the work of [24] We Consider the following notations to model the problem constraints:

- H_{min_i}: is the minimum time interval between two vehicles of the same line in station 'i'.
- H_{max_i}: is the maximum time interval between two vehicles of the same line in station 'i'.
- t_{ij}: $t_j - t_i$ is the elapsed time between the departure time t_j of station 'j' and the departure time t_i of station 'i'. 'i' and 'j' represent the two successive stations of the link l_{ij} respectively.
- T_{ce_i}: is the estimated total travel time of trip 'i'.
- T_{ct_i}: is the scheduled total travel time of trip 'i'.
- N_i: is the number of trips conducted at station i
- I_{PUN_i}: is the punctuality index at station 'i'.
- $I_{PUN_{max}}$: is the max punctuality index allowed at station 'i'.
- I_{REG_i}: is the regularity index at station 'i'.
- $I_{REG_{max}}$: is the max regularity index allowed at station 'i'.

The problem is infeasible unless the following constraints are satisfied for each trip:

$$I_{REG_i} \le I_{REG_{max}} \tag{9}$$

$$I_{PON_i} \leq \min(I_{PON_{max}}, I_{REG_{max}}) \tag{10}$$

$$t_i \leq N_i.H_{max_i} \tag{11}$$

$$t_i \geq (N_i - 1).H_{min_i} \tag{12}$$

$$T_{ce\_i} \leq T_{cmax}, T_{cmax} = T_{ce\_t} + \left(n * I_{REG_{max}}\right) \tag{13}$$

$$t_i \in [0, I_{REG_{max}}] \tag{14}$$

These constraints are mandatory to verify the following situations:

- Not to exceed the maximum regularity limit allowed (Eq. 9)
- Not to catch up with the regulated trip (Eq. 10)
- Not to exceed the maximum time allowed during a regulation. (Eq. 11)
- Respect the minimum regularity between vehicles of the same line (Eq. 12)
- Not to exceed the maximum time allowed for a given trip (Eq. 13)
- Not to have a conjunction of two consecutive trips in the starting station (Eq. 14)

6   Control System Modeling

6.1   Multi-agents Model

We will implement a control system based on a multi-agent approach. The agents in our system are heterogeneous, and their interactions reflect the network state of public transport. To guarantee our goal, the system must detect and manage perturbations by providing a good coordination between the agents. Each agent has a specific role in its environment. The agents in our model are described as follows:

Vehicle. The vehicle agent memorizes all the data that characterizes it. For example, position, type, speed, capacity, number of passengers, line assigned, etc. It collects the data related to the current link and the values of the KPIs. It calculates the overall performance to find the value of the performance variation to detect perturbation. In case of disturbance, it transmits a call to the regulator to trigger the decision-making step. Also, each vehicle agent transmits, regularly, its properties with those of the current link to the arrival station agent to estimate the remaining time.

Link. It represents the transition between two successive stations. It should be linked to at least one line. It stores two types of information (i) Static properties: distance, maximum speed allowed and maximum density. (ii) Dynamic properties: average speed, current density. This data is sent to the vehicle agent. The link agent used to analyse and detect link congestion by calculating the speed performance index as an indicator to evaluate the traffic condition of the connection. This indicator is passed to the KPI agent to calculate its value.

Station. The station agent is linked to one or more lines. Each agent memorizes the passenger arrival and departure flows, as well as the scheduled and actual passage times of the vehicles. It receives the necessary vehicle properties to calculate the remaining arrival time. Then it gives the necessary data to the KPI agents (scheduled arrival time, remaining time) so that they can measure the performance of the vehicle.

KPI. It calculates the value of the key performance indicator and transmits it to the concerned vehicle agent. In our system the KPIs are classified by objectives.

Regulator. Each "regulator" agent is responsible for a geographical area of the network. It receives the KPIs of each vehicle in perturbation. Then it performs an optimization to find the best regulation action.

6.2 Inter-agent Message Exchange

The system has to detect and solve traffic disturbances. It consists of a company of agents. These agents communicate with each other via messages. To guarantee our.
objective, the system must detect and manage disturbances by providing good coordination between the agents. Each agent has a specific role in its environment.
The process starts with an analysis of the traffic situation. Each vehicle receives the properties of the current link with the calculated speedometer. Then these properties are used to estimate the delay time with the coordination of the arrival station. In case of a disturbance, the KPI agent calculates the value of its criterion and sends it to the relevant (zone) regulator. In turn, the "regulator" agent tries to find the control action as a result

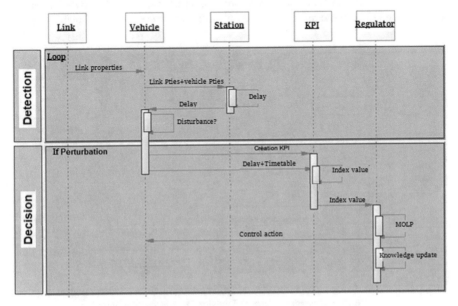

Fig. 6. Dynamic view of our system.

of an optimization solution using the MOLP method described above. Finally, he gives the order to each vehicle concerned to apply the chosen action.

It should be noted that after each implementation of a control action, each agent must update his knowledge base with any relevant changes in the current traffic situation. Figure 6 presents a dynamic view of our system. It shows the sequence diagram of messages exchanged between agents.

7   Computational Results

7.1   Simulation Model

In this section, problems that are taken from our previous experimentation [1] are solved by proposed fuzzy solution approach to determine the indicator weights using linear membership functions of our MOLP presented in Sect. 5.

In this experiment, we test the control strategy of our system on a simulator of a real network in the city of Portland in Oregon (See Fig. 7). The data was collected from the general transit department of the District of Oregon's "Tri-County Metropolitan Transportation" (TriMet) and imported into AnyLogic as GTFS files to model the map data of the TriMet network. We test our control system model on the "2 Division" line connecting Portland City Center and Gresham Transit Center (round trip). This line has eight stations with 86 outbound trips from 5:26 AM to 1:41 AM of the next day and 87 return trips from 4:09 AM to 12:42 AM of the next day, as well as connections to several lines (https://ride.trimet.org/).

Our system simulation model includes a graphical interface that visualizes the inputs and outputs of the simulation. The network infrastructure and stations are displayed graphically, and the vehicle movements are animated. Our model allows loading of different scenarios, provides a geographical traffic map, and zoom, as well as pause.

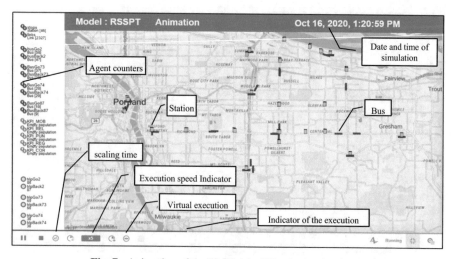

Fig. 7. Animation of the "2-Division" line in the simulator.

The simulation also provides the numerical data result in sheet and chart resolution. The simulation animation shown in Fig. 8 confirms that the route and station layouts in the map correspond to reality. It also shows precisely that the movement of the vehicles in their real trips.

7.2 Scenario and Results

The scenario presents traffic congestion observed on the "2-Division" line due to bad weather conditions caused by fog. It occurred in the morning on the 10th trip at stop #1375 (SE Division & 12th). Figure 8 shows the alert signal for the disruption with the regulation action held by TriMet. The chosen solution indicates that the service in the station is temporarily disrupted, and passengers are advised to go to the nearby station at address 2314. There is no action applied to the vehicle.

It is the same senario of [1] solved by using our fuzzy logic approach "Objective Average Variations" (OAV) method to determine the weigh of each separate objective. The average degrees of objective functions are found as $Avg_{\mu(I_{PUN})} = 0.74$, $Avg_{\mu(I_{REG})} = 0.69$ and $Avg_{\mu(I_{COR})} = 0.82$. The values of objectives are found $I_{PUN} = 0.34$, $I_{REG} = 0.58$ and $I_{COR} = 0.17$; Table 3 shows the results of weights with"OAV" method.

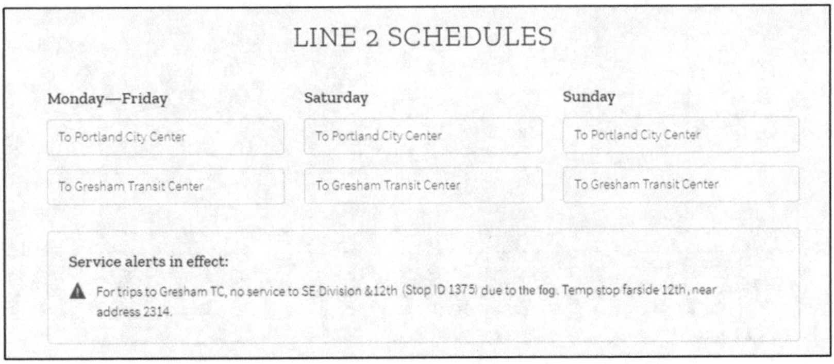

Fig. 8. Disruption alert in the traffic of the itinerary "2 Division".

Table 3. Weight distribution of the indices.

$w_{I_{PUN}}$	$w_{I_{REG}}$	$w_{I_{COR}}$
0.41	0.34	0.25

In Fig. 10, the contribution of our control system is shown by the considerable reduction of the delays of the disturbed trips comparing those of the Fig. 9.. It also shows that our system detects the disturbance earlier and the resolution of the disturbance is faster in both cases of disturbances when our control system is applied. In fact, the disturbance was detected in trip 5 instead of trip 7 and it was fully resolved on trip 12 instead of trip 15.

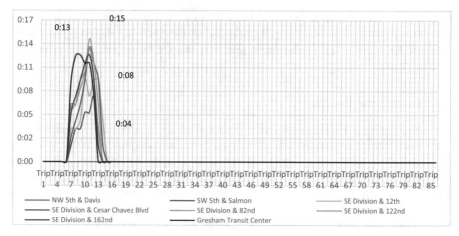

Fig. 9. Observed delays on each station using our previous control system.

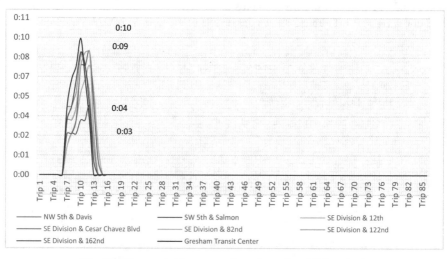

Fig. 10. Observed delays on each station using OAV method.

In Fig. 11, we plot the average of F values calculated on each trip of the itinerary "2-Division" with the current control system of TriMet and with our previous and new control system with OAV method. The results confirm an improvement in the quality of service by minimizing the values of the variation F during the disturbance comparing to the two other cotrols systems.(current of TriMet and our previous). The area separating the two curves represents the gain in performance variation when using OAV method as a technic to determine the weight.

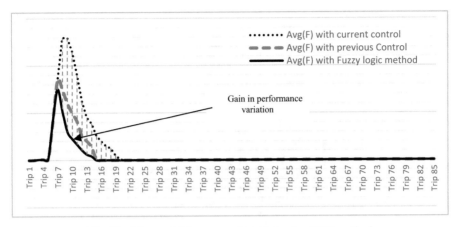

Fig. 11. Observed delays on each station using OAV method.

We present in the Figs. 12 and 13 the percentage increase of passengers waiting by station (PI) on each trip for both disturbances. This percentage is relatively proportional to the bus delays. We used the Poisson distribution to simulate the number of passengers at station because both phenomena arriving and departing the stations of passengers are independents. The simulation shows the contribution of OAV method by minimizing the PI value on the stations.

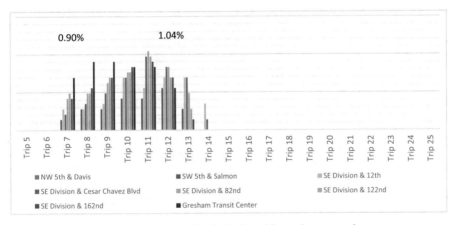

Fig. 12. PI per station on disturbed trips with previous control system.

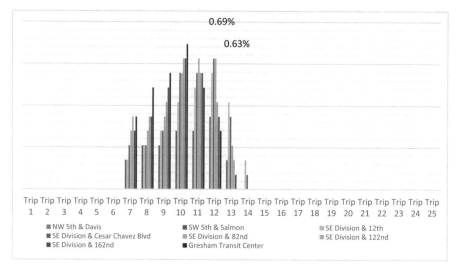

Fig. 13. PI per station on disturbed trips using OAV method.

8 Conclusion and Future Work

The main objective of this study was to model and develop a transit control system. This system simulates and controls the traffic of a transit system. It detects in real time the traffic disturbances of the itineraries and generates the appropriate control action. The modeling of the system is based on a multi-agent approach dealing with an optimization problem. The mathematical model of the optimization problem is based on key performance indicators (KPIs) related to the traffic efficiency of public transit.

An OAV method was developed to determine the weights of the KPIs in the formulation of the mathematical model. This method takes into account the influence of the KPIs on the determination of the control action. It presents a considerable contribution compared to the current control system of the transit compagnie and of those of our previous work.

To validate our control system, a simulation model reflecting the real transit has been built. The development is done with AnyLogic. It is an agent-based modeling simulator. The simulation of our control system gives a visual and mathematical results justifying the choice of the control action. The emergence of the agents' interactions of our model after simulation give good results.

The experiments are performed on real transit system. Based on real scenarios, we show that the proposed model is able to detect and manage disturbances with a better performance compared with the results obtained in our previous experimentation.

Finally, in a perspective toward optimization domain, we can enrich the work by adding methods related to the determination of weigths. A relevant recent method are: Asynchronous Fault Detection for Interval Type-2 Fuzzy Nonhomogeneous Higher-level Markov Jump Systems with Uncertain Transition Probabilities.

In the multi-agent modeling we mention two tracks: (i) to be able to manage disturbances in unfamiliar situations (unknown disturbance, new traffic parameter, etc.), we need to improve the behaviour of the system by providing an evolutionary approach in its resolution. This approach consists of making sure that, thanks to the regulator agent, the system can remember the results for these types of situations. The model should then suggest a fast neighborhood solution as a future action with new experiences and update the regulator's knowledge base by inserting these new rules to cope with future situations. (ii) to orient the control system towards the operator, we need to change the goal and include other performance measures related to the costs of transport. This track consists of adapting the optimization method to resolve a problem with antagonistic variables. Variables directed towards the user's view and variables directed towards the operator's view.

References

1. Morri, N., Hadouaj, S., Ben Said, L.: Agent-based intelligent KPIs optimization of public transit control system. In Proceedings of the 18th International Conference on Informatics in Control, Automation and Robotics (ICINCO 2021), pp. 224–231 (2021). ISBN: 978-989-758-522-7. https://doi.org/10.5220/0010616302240231
2. Kagho, G.O., Balac, M., Axhausen, K.W.: Agent-based models in transport planning: current state, issues, and expectations. Procedia Comput. Sci. **170**, 726–732 (2020)
3. Cré, I., Rupprecht, S., Bührmann, S.: The development of local implementation scenarios for innovative urban transport concepts: the NICHES+ approach. Procedia Soc. Behav. Sci. **48**, 1324–1335 (2012)
4. Haugeneder, H., Steiner, D.:. A multi-agent approach to cooperation in urban traffic. In: CKBS-SIG (Conference: 1993: Keele, England). CKBS-SIG Proceedings (1993). https://doi.org/10.5220/0010616302240231
5. Wooldridge, M., Müller, J.P., Tambe, M. (eds.): ATAL 1995. LNCS, vol. 1037. Springer, Heidelberg (1996). https://doi.org/10.1007/3-540-60805-2
6. Ossowski, S., et al.: Multi-agent systems for decision support: a case study in the transportation management domain. Appl. Artif. Intell. **18**(9–10), 779–795 (2004)
7. Nagel, K., Rickert, M., Barrett, C.L.: Large scale tra c simulations. In: Proceedings of VECPAR, vol. 96 (1997)
8. Raney, B., Nagel, K.: 3.5 an improved framework for large-scale multi-agent simulations of travel behaviour. In: Towards Better Performing Transport Networks, vol. 34, p. 305 (2006)
9. Adler, J.L., Satapathy, G., Manikonda, V., Bowles, B., Blue, V.J.: A multi-agent approach to cooperative traffic management and route guidance. Transp. Res. Part B: Methodological **39**(4), 297–318 (2005)
10. Davidsson, P., Henesey, L., Ramstedt, L., Törnquist, J., Wernstedt, F.: An analysis of agent-based approaches to transport logistics. Transp. Res. Part C: Emerg. Technol. **13**(4), 255–271 (2005)
11. Dia, H.: An agent-based approach to modelling driver route choice behaviour under the influence of real-time information. Transp. Res. Part C: Emerg. Technol. **10**(5–6), 331–349 (2002)
12. Nagel, K., Marchal, F.: Computational methods for multi-agent simulations of travel behavior. In: Proceedings of International Association for Travel Behavior Research (IATBR), Lucerne, Switzerland (2003)
13. Dia, H., Purchase, H.: Modelling the impacts of advanced traveller information systems using intellingent agents. Road Transp. Res. **8**(3), 68 (1999)
14. Ferber, J.: Les systèmes multi-agents: un aperçu général. Techniques et sciences informatiques **16**(8) (1997)

15. Tromp, M., Liu, X., Graham, D.J.: Development of key performance indicator to compare regularity of service between urban bus operators. Transp. Res. Rec. J. Transp. Res. Board 2216, 33–41 (2011)
16. M. Napiah et al., (2015) M. Napiah,, I. Kamaruddin and Suwardo, "Punctuality index and expected average waiting time of stage buses in mixed traffic", WIT Transactions on The Built Environment, Vol 116, © 2011 WIT Press. ISSN 1743–3509 (on-line)
17. Cats, O., Burghout, W., Toledo, T., Koutsopoulos, H.N.: Mesoscopic modeling of bus public transportation. In: No. 2188, Transportation Research Board of the National Academies, Washington, D.C., 2010, pp. 9–18 (2010)
18. Bhouri, N., Aron, M., Scemama, G.: Gini Index for Evaluating Bus Reliability Performances for Operators and Riders. Transportation Research Board, Washington, p. 13 (2016)
19. Carosi, S., Gualandi, S., Malucelli, F., Tresoldi, E.: Delay management in public transportation: service regularity issues and crew re-scheduling. In: 18th Euro Working Group on Transportation, EWGT, Delft, The Netherlands(2015)
20. Noorfakhriah, Y., Madzlan, N.: Public transport: punctuality index for bus operation. World Acad. Sci. Eng. Technol. Int. J. Civil Environ. Eng. 5(12) (2011)
21. Chen, X., Yu, L., Zhang, Y., Guo, J.: Analyzing urban bus service reliability at the stop, route, and network levels. Transp. Res. Part A 43, 722–734 (2009)
22. Vaniyapurackal, J.J.: Punctuality index for the city bus service. Int. J. Eng. Res. 4(4), 206–208 (2015). ISSN:2319–6890
23. Saberi, M., Zockaie, K.A.: Definition and properties of alternative bus service reliability measures at the stop level. J. Public Transp. 16(1), 97–122 (2013)
24. Ceder, A.: Public Transit Planning and Operation: Theory, modelling and practice. Elsevier Ltd., Amsterdam (2007)
25. Ma, Z., Ferreira, L., Mesbah, M.: A framework for the development of bus service reliability measures. In: Australasian Transport Research Forum, Brisbane, Australia (2013)
26. Dueker, K.J., Kimpel, T.J., Strathman, J.G.: Determinants of bus dwell time. J. Public Transp. (2004)
27. Levinson, H.: Analyzing transit travel time performance. Transp. Res. Rec. 915 (1983)
28. Tran, V.T., Eklund, P., Cook, C.: Toward real-time decision making for bus service reliability. In: International Symposium on Communications and Information Technologies(2012)
29. Cats, N., Koutsopoulos, H., Burghout, W.: Impacts of holding control strategies on transit performance: a bus simulation model analysis. Transp. Res. Rec. J. Transp. Res. Board 216, 51–58 (2011)
30. Cambridge Systematics Inc., PB Consult Inc., and System Metrics Group. Analytical tools for asset management. 545, NCHRP report, 2005(2005)
31. European Commission. White paper - Roadmap to a single European Transport Area - Towards a competitive and resource efficient transport system (2011)
32. Yan, X.Y., Crookes, R.J.: Reduction potentials of energy demand and GHG emissions in China's road transport sector. Energy Policy 37, 658–668 (2009)
33. Van Oort, N., van Nes, R.: Control of public transport operations to improve reliability : theory and practice. Transp. Res. Rec. 2112 (2009)
34. Heylighen, F., Joslyn, C.: Second Order Cybernetics. Principia Cybernetica2001
35. Hahn, J., Edgar, T.F.: An improved method for nonlinear model reduction using balancing of empirical gramians. Comput. Chem. Eng. 26(10), 1379–1397 (2002)
36. Ashby, W.R.: Automata Studies: Annals of Mathematics Studies, No. 34. Princeton University Press, Princeton (1956)
37. Zaidi, K.F., Agrawal, N.: Microstencil-based spatial immobilization of individual cells for single cell analysis. Biomicrofluidics 12(6), 064104 (2018)
38. Naaz, S., Alam, A., Biswas, R.: Effect of different defuzzification methods in a fuzzy based load balancing application. Int. J. Comput. Sci. Issues (IJCSI) 8(5), 261 (2011)

Robotics and Automation

A Two-Stage Trajectory Prediction Algorithm for Mobile Robots Combining the Bayesian and the DMOC Frameworks

Wei Luo and Peter Eberhard[✉]

Institute of Engineering and Computational Mechanics, University of Stuttgart,
Pfaffenwaldring 9, 70569 Stuttgart, Germany
{wei.luo,peter.eberhard}@itm.uni-stuttgart.de

Abstract. In many real-world robotic applications, predicting an observed agent's trajectory can significantly improve cooperation efficiency, avoid potential collisions, and alleviate the communication strain in multi-agent scenarios. Given a well predicted trajectory of the observed other agents, the observer can either establish a clear path to its destination or collaborate with the observed agent in a more energy-efficient manner. To address this challenge, in this paper, a two-stage trajectory prediction algorithm is proposed based on the observed agent's previous trajectory data. First, the potential destination of the observed robot is guessed, and the future guessed path is then sampled using the Monte-Carlo sampling approach within a Bayesian framework. Then, an optimization problem based on a discrete mechanics and optimal control (DMOC) framework with complementarity constraints is proposed to forecast a more reasonable trajectory, while the previously predicted path is used as the reference. Finally, several experiments are undertaken to verify the performance of the proposed algorithm in simulations and real-world applications with our holonomic and nonholonomic mobile robots.

Keywords: Intention evaluation · Monte-Carlo sampling · Optimization · Trajectory prediction · Complementarity constraint · Mobile robot · Discrete mechanics and optimal control

1 Introduction

Mobile robots are currently among the most popular robotic devices in academic and industrial applications, including object transports [10], target search [16], and medical rescues [24]. As the number of involved mobile robots increases, operation and cooperation in a multi-robot scenario or even combined with human beings become challenging issues, as mobile robots are commonly operated independently in real-world applications, and safety is critical. In such

O. Gusikhin et al. (Eds.): ICINCO 2021, LNEE 1006, pp. 39–62, 2023.
https://doi.org/10.1007/978-3-031-26474-0_3

cases, robots must be able to predict the likely trajectory of agents of inter-est, i.e. their intention, within the same environment based on the information of the scenario and the observation data collected via onboard sensors. Given the predicted future trajectory of the observed agent, one can plan a collision-free course through the scenario, or catch up with other robots through an efficient approach.

Let us begin with a hypothetical scenario for an investigation. A flying quadrotor is observing a scene with its onboard sensors. Meanwhile, one or sev-eral mobile robots are moving across the scene with their own intentions, which the quadrotor is unaware of, see Fig. 1(a). The quadrotor wants to approach and contact one of these mobile robots. However, this study will not discuss the control of the quadrotor or its approach path. Rather than that, we focus on predicting the unknown but most likely future trajectory of the quadrotor's approaching target robot, just based on information about the observed mobile robot's past motion. Note that this has to assume that the mobile robot behavior is 'reasonable' and has a certain intention that should be predicted (instead of, e.g., just performing random walks). For instance, in Fig. 1(b), we assume that the observed mobile robot will progress to one of the exits as its destination. On the other hand, a pure random path would not follow an intention so that no prediction would be possible. Thus, in this work, we presume that the observed mobile robot has its own intention, despite the fact that the quadrotor has no access to or direct control over its motion or trajectory planning. Moreover, the mobile robot is considered to take a reasonable short path to its destination instead of roaming in the scenario.

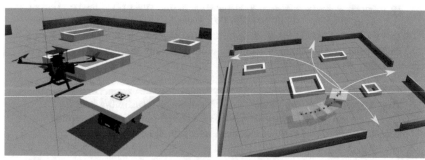

(a) A flying quadrotor with its observing mo-bile robot
(b) Potential trajectories to be predicted for the mobile robot

Fig. 1. A potential investigation scenario.

To address the mentioned issue, in this work, we present a two-stage app-roach for predicting the observed mobile robot's future trajectory based on its past observable information in a known scene. The first stage involves evaluat-ing the intention of the observed mobile robot and estimating its potential path waypoints using the proposed Bayesian framework. Then, in the second stage, the estimated path waypoints are utilized as the reference in an optimization problem. By solving the proposed optimization problem with observable constraints, an even more reasonable trajectory can be obtained to represent the potential future behavior of the observed mobile robot.

1.1   Related Works

To predict the future trajectory, several researchers have defined a prior model to characterize the agent's motion of interest. In [22], it is shown that a constant velocity model could outperform the state-of-the-art models. Furthermore, one used a polynomial model to approximate the agent's future trajectory [23]. Besides, neural network-based approaches have also demonstrated remarkable achievements in predicting the trajectory of an observed agent based on the collected previous data. In [11], the generative adversarial networks (GANs) were applied to predict pedestrian's trajectory. Moreover, the long short-term memory (LSTM), which is a variation of recurrent neural networks (RNNs), was used to predict the trajectory of vehicles on the street [4]. More recently, transformer-based approaches have also shown their superiority in handling time-series predictions [9, 26].

Moreover, instead of directly predicting the observed agent's trajectory, some approaches conducted their predictions under goal conditions. The procedure of these approaches can usually be divided into two or more stages. In the first stage, the goal of the observed agent is evaluated, and subsequent stages of prediction will then take into account the evaluated a priori knowledge. In [2], a Bayesian formulation was used to evaluate the agent's travel goal, which was represented with a probability distribution and used to estimate the future trajectory of the observed agent recursively. In [5], the goal condition was evaluated in a GAN framework, and the proposed method showed a better performance compared to other state-of-the-art approaches on some classical pedestrian datasets.

1.2   Contributions

Our proposed algorithm has a two-stage structure: a Bayesian framework that predicts on the basis of the guessed intention the potential path waypoints and a trajectory predictor that solves the proposed optimization problem, which takes the previously predicted path waypoints as the reference. The Bayesian framework is inspired by the research in [2], but our proposed probability dynamics function considers not only the change in travel distance to the potential destination but also its current motion tendency which has significantly improved the sampling efficiency. The second stage involves an optimization problem for estimating an even better trajectory for the observed robot. In contrast to our prior work [17], we employ the variational approach in the proposed optimization problem, which is based on discrete mechanics and optimal control (DMOC) [20], and incorporates complementarity constraints. To our best knowledge, this is the first implementation based on a DMOC framework that allows for the use of complementarity constraints and solves the trajectory prediction problem.

Additionally, two typical mobile robots, namely holonomic and nonholonomic mobile robots, are investigated in the simulation experiment and later in the real-world experiment with our self-designed mobile robots. In all experiments, our proposed algorithm can predict a reasonable trajectory for the investigated mobile robots, given their past trajectories.

This paper is organized as follows. In Sect. 2, the Bayesian framework is introduced, which comprises the intention estimation and a motion tendency-based intention probability function. Subsequently, in Sect. 3 the proposed optimization approach based on DMOC in the second stage is proposed. Then, in Sect. 4, both simulation and hardware experimental results with our holonomic and non-holonomic mobile robots are illustrated. Finally, the conclusions and an outlook to future work are given in Sect. 5.

Notation. This paper uses lowercase for scalars, lowercase bold for vectors, and uppercase bold for matrices. Furthermore, it is anticipated that the past trajectory of the observed robot is available from measurements and will be utilized as input of the proposed algorithm of this work. At each time step t_i, the past observation set is indicated by $\boldsymbol{X}_{1:i} := \{\boldsymbol{x}_j = (x_j, y_j, \psi_j) \in \mathbb{R}^3 | j = 1, \ldots, i\} \in \mathscr{X}$ and a sequence of the predicted robot's future poses is denoted by $\check{\boldsymbol{X}}_{1:n_\mathrm{p}} := \{\boldsymbol{x}_j = (x_j, y_j, \psi_j) \in \mathbb{R}^3 | j = 1, \ldots, n_\mathrm{p}\} \in \mathscr{X}$ for the subsequent time steps, where $\mathscr{X}$ denotes the continuous space of the whole scenario, and the prediction horizon is specified as n_p. Note that without any special notification, the heading of the observed mobile robot $\psi_{(\cdot)}$ is neglected in the first stage since we utilize only the positions of the observed robot and the predicted path contains only the geometrical connections. Besides, the positions of the robot can also be represented in a discrete roadmap, which is denoted by $\hat{\mathscr{X}}$ in this work.

2   Intention Evaluation-Based Bayesian Path Waypoint Prediction

The problem of predicting the next possible path waypoint $\boldsymbol{x}_{i+1}$ for the observed mobile robot can be reduced to determining its most likely future position given its prior path waypoints $\boldsymbol{X}_{1:i}$, which can be further represented by a probability function $\Pr(\boldsymbol{x}_{i+1}|\boldsymbol{X}_{1:i})$. Furthermore, in previous work [2,5], the intention of the observed agent assists the prediction process with great effort. Thus, in this work, we also assume that each observed robot should have its own intended destination instead of just wandering in the scenario, and we want to find a possible connection between its current position and the probable destination. Although the precise motion destination of the observed robot cannot be known in advance or be planned by the observer, we can generally assume that the set of all potential destinations, such as locations for loading/unloading of goods, charging stations or parking spots, is feasible and finite, as denoted by $\boldsymbol{\Theta} := \{\theta_\eta | \eta \in [1, \ldots, n_t]\}$, where the number of the potential destinations is defined as n_t.

Additionally, to guess the probability distribution of the potential destinations of the observed robot, the sequence of the path waypoints at individual time steps to the potential destination can represent a directed probabilistic graphical model, as illustrated in Fig. 2, where the parameter θ indicates an arbitrary destination that is considered as a latent variable in the probabilistic graphical model. Based on the description above, the probability from the current position of the observed robot $\boldsymbol{x}_i$ to a potential destination θ_η given the previously observed data $\boldsymbol{X}_{1:i}$ can be determined by

$$\Pr(\boldsymbol{x}_{i+1}|\boldsymbol{X}_{1:i}, \theta = \theta_\eta) = \underbrace{\Pr(\boldsymbol{x}_{i+1}|\boldsymbol{x}_i, \theta_\eta)}_{\Pr_1} \times \underbrace{\Pr(\theta_\eta|\boldsymbol{X}_{1:i})}_{\Pr_2}, \tag{1}$$

where the probability function $\Pr_1$ indicates the probability from the current position to a potential destination θ_η that the position $\boldsymbol{x}_{i+1}$ at the next time step will be chosen. In general, given the probabilistic graphical model in Fig. 2, the joint probability of the total model until time step $i + 1$ can be obtained by

$$\Pr(\boldsymbol{x}_{i+1}, \boldsymbol{x}_i, \dots, \boldsymbol{x}_1, \theta) = \Pr(\theta) \prod_{j}^{i+1} \Pr(\boldsymbol{x}_{j+1}|\boldsymbol{x}_j, \theta). \tag{2}$$

Based on the Bayes' theorem, the probability function is

$$\Pr(\boldsymbol{x}_{i+1}|\boldsymbol{x}_i, \boldsymbol{x}_{i-1}, \dots, \boldsymbol{x}_1, \theta) = \frac{\Pr(\boldsymbol{x}_{i+1}, \boldsymbol{x}_i, \dots, \boldsymbol{x}_1, \theta)}{\Pr(\boldsymbol{x}_i, \boldsymbol{x}_{i-1}, \dots, \boldsymbol{x}_1, \theta)} = \Pr(\boldsymbol{x}_{i+1}|\boldsymbol{x}_i, \theta), \tag{3}$$

which also obeys the Markov chain rule.

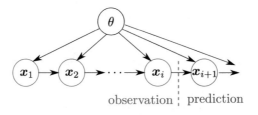

Fig. 2. Illustration of the probabilistic graphical model.

Moreover, the probability function $\Pr_2$ describes the posterior guess of the mobile robot's intention given its previously observed trajectory. Both probability functions $\Pr_1$ and $\Pr_2$ will be specified in the following subsections.

Furthermore, for an effective prediction, the potential path waypoints in a discrete space $\hat{\mathscr{X}}$ instead of in a continuous space $\mathscr{X}$ will be searches in the first stage of the proposed algorithm. In this work, several randomly dispersed nodes are produced in advance in the scenario composing a discrete space representing the potential visited path waypoints maybe later traversed by the observed mobile robot. Moreover, inspired by motion planning algorithms, the k-nearest optimal probabilistic roadmap (k-PRM*) is utilized in this work to create a roadmap [14] that results in an undirected graph with collision-free connections between the previously generated nodes in the scenario. Two essential parameters, which concern the probabilistic completeness of the generated roadmap, are specified by

$$
\begin{aligned}
r_{\mathrm{PRM}} &= (\sqrt{6}(A_{\text{free-space}}/\pi)^{0.5} + 1)(\log(n_{\text{nodes}})/n_{\text{nodes}})^{0.5}, \\
k_{\mathrm{PRM}} &= 2e \log(n_{\text{nodes}}),
\end{aligned} \tag{4}
$$

where the free area of the scenario is marked with $A_{\text{free-space}}$ [m^2], and the total number of nodes on the roadmap is denoted by n_{nodes}. Furthermore, to improve the real-time computing performance, the shortest paths between nodes are determined by the A* algorithm in advance, and their corresponding path distances will be calculated offline and stored in a query database.

2.1 Intention Evaluation

In this work, the intention of the observed robot is guessed by computing a probability for each destination based on its previously observed trajectory, and it will be later represented with a probability function. Two factors are tracked in the proposed intention probability function. The first factor evaluates the path reduction when moving to a certain destination θ_η given the past data. Compared to [2], we only take the l_o latest robot positions since historical observations may hinder the response to the intention estimation. Additionally, the prior motion direction of the observed robot may also indicate the likely destination, following the intuitive assumption that the motion direction quite likely somehow heads in the direction of the destination. The motion tendency-based intention probability function is built in light of these two factors as

$$
\Pr(\theta_\eta | \boldsymbol{X}_{1:i}) \propto \prod_{j=i-l_o}^{i} \exp(f_{\mathrm{d}}(\boldsymbol{x}_{j-1}, \theta_\eta) - f_{\mathrm{d}}(\boldsymbol{x}_j, \theta_\eta)) \underbrace{(f_{\mathrm{r}}(\boldsymbol{x}_{j-1}, \boldsymbol{x}_j, \theta_\eta) + 1)}_{\text{motion tendency}}, \tag{5}
$$

where the function f_{d} indicates the shortest path distance between two specified positions. It is worth noting that the shortest path distance does not always correspond to the shortest Euclidean distance, as there may be no direct connection between these two positions in the roadmap due to missing waypoints there. The function f_{r} returns the cosine of the angle between three given positions in 2D, which is defined as

$$
f_{\mathrm{r}}(\boldsymbol{a}, \boldsymbol{b}, \boldsymbol{c}) = \frac{(\boldsymbol{b} - \boldsymbol{a}) \cdot (\boldsymbol{c} - \boldsymbol{b})}{\|\boldsymbol{b} - \boldsymbol{a}\|\|\boldsymbol{c} - \boldsymbol{b}\|}. \tag{6}
$$

2.2 Improved Probabilistic Dynamics Model

In the first stage, to predict the potential motion of the observed robot, the dynamics of the observed robot is described with a probabilistic dynamics model that estimates the probability of a candidate in the roadmap given its previous position and the potential destinations. Instead of focusing exclusively on the path distance (also called the geodesic distance) in [2], we introduce a new parameter β to include the mechanical effect of linear momentum on the probability function. The improved probabilistic dynamics model is defined as

$$\Pr(\boldsymbol{x}_{i+1}|\boldsymbol{x}_i, \theta_\eta) \propto \exp(-\alpha(f_\mathrm{d}(\boldsymbol{x}_i, \boldsymbol{x}_{i+1}) + f_\mathrm{d}(\boldsymbol{x}_{i+1}, \theta_\eta) - f_\mathrm{d}(\boldsymbol{x}_i, \theta_\eta)))\beta, \qquad (7)$$

where the parameter α is non-negative and selected by the user. When $\alpha \to 0$, the probability for each node in $\tilde{\mathcal{X}}_i$ is nearly equal when the effect of parameter β is ignored. On the contrary, if $\alpha \to +\infty$, it is assumed that the robot always follows the shortest path to its destination. Hence, choosing the parameter α could be difficult, as it is challenging to balance the exploration and the exploitation when considering only the distance relationship with parameter α, especially in the scenarios with multiple destinations. The parameter β is given by

$$\beta = \max\{1\mathrm{e}{-6}, f_\mathrm{r}(\boldsymbol{x}_{i-1}, \boldsymbol{x}_i, \tilde{\boldsymbol{x}}_{i+1})\|\boldsymbol{x}_i - \boldsymbol{x}_{i-1}\|\}, \qquad (8)$$

where $\tilde{\boldsymbol{x}}_{i+1}$ denotes a candidate for the next position.

Figure 3 depicts an example for the estimation with the improved probabilistic dynamics model. Note that it is not wise to calculate the probability for every node in the roadmap. Rather than that, only a cluster in close proximity to the mobile robot's current position is examined. In this work, given the definition of the PRM* roadmap, we include only those nodes in the calculation that have a direct connection with the nearest node to the mobile robot's current position, forming a space marked with $\tilde{\mathcal{X}}_i \in \hat{\mathcal{X}}$. Given the past course $\boldsymbol{X}_{i-2:i}$ indicated by red circles, the candidates located in the shadow region are less likely to be picked as the potential path waypoints of the mobile robot according to the proposed probabilistic dynamics model in Eq. (7).

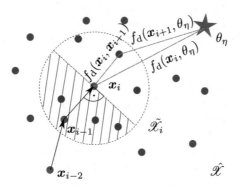

Fig. 3. Illustration for the probabilistic dynamics model. The dotted circle illustrates the candidate area $\tilde{\mathcal{X}}_i$, and the nodes that the robot can arrive within next time step, are marked with brown circles. The nodes which are out of range, are marked with blue circles. The green dashed lines indicate shortest path connection between two connected nodes. The red star represents a potential destination in the scenario.

2.3   Bayesian Waypoint Prediction Framework

Intuitively, based on Eq. (1) one can recursively predict the probability of the next waypoint at time step $i + j$ with $j \geq 0$, given a fixed posterior intention evaluation of the mobile robot in Eq. (5), since there are no more possible observations to evaluate the destination up the time step i. However, as mentioned in [2], the analytical evaluation is difficult due to the branching factor of the roadmap. As a result, the trajectory waypoints will be estimated through the Monte-Carlo sampling approach.

Based on the evaluated intention probability distribution in Eq. (5), $N_\eta = N \times \Pr(\theta_\eta | \boldsymbol{X}_{1:i})$ samplings will be executed from current position $\boldsymbol{x}_i$ to a candidate destination θ_η. Each sampling chooses the next feasible position node $\hat{\boldsymbol{x}}_{i+j}$ in the region $\tilde{\boldsymbol{X}}_{i+j-1}$ based on the probability implied by the improved probabilistic dynamic model in Eq. (7), and the results of each prediction step j are organized accordingly. The direct sampling results for four individual steps are illustrated in Fig. 4. Note that the prediction horizon is constrained in order to get a balance between the computational time and the exploration.

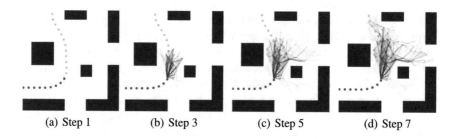

| (a) Step 1 | (b) Step 3 | (c) Step 5 | (d) Step 7 |

Fig. 4. Direct Monte-Carlo sampling results.

To acquire the final predicted path following the Monte-Carlo sampling, the proposed approach employs a sum-pooling procedure rather than greedily choosing the most sampled nodes at each prediction time step. Each node visited in $\hat{\boldsymbol{X}}_{i+1:i+j}$ at the time step j will be pooled and converted into a grid graph, as illustrated in Fig. 5(b). Then, the visiting times of each grid on the graph are summed up to create a counting map $\boldsymbol{M}_c$ for each prediction time step, which indicates the grid's probability of being visited. While the sum-pooling process sacrifices the prediction accuracy, it also reduces the unbalanced distribution of the generated nodes on the roadmap $\hat{\mathscr{X}}$ and avoids the prediction of excessively short paths, which is especially true when a relatively small α in Eq. (7) is chosen. Finally, the most frequently visited grid to the goal region at each prediction time step will be recorded and expressed as the predicted waypoint of the observed robot as $\boldsymbol{X}_{1:n_{\mathrm{w}}}^{\mathrm{pred}} := \{\boldsymbol{x}_j^{\mathrm{pred}} \in \mathbb{R}^2 | j \in [1, \ldots, n_{\mathrm{w}}]\}$. In the illustration in Fig. 5(c), the connected gray cells represent the final prediction result. Compared to the greedy strategy result in Fig. 5(a), the predicted trajectory following the sum-pooling procedure is smoother and more reasonable as the prediction for the observed mobile robot since some unphysical connections are avoided.

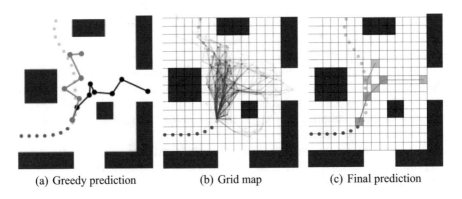

| (a) Greedy prediction | (b) Grid map | (c) Final prediction |

Fig. 5. Sum-pooling procedure.

3   Trajectory Prediction Based on the DMOC Framework

After the sum-pooling procedure, a sequence of the predicted path waypoints is selected for the observed mobile robot. However, this is inappropriate to be utilized directly in most robotics applications, as it only provides a vague hint about the future paths and contains only geometrical connections in the roadmap. There are two apparent insufficiencies. On the one hand, the distribution of the path waypoints is highly dependent on the generated roadmap. The physical behavior of the observed robot is disregarded, since the estimated path is only represented with straight lines between the estimated path waypoints. On the other hand, the time from point to point is not specified and thus velocities and accelerations are not defined.

To address the insufficiencies above, in the second stage an optimization problem is formulated that uses the previously predicted path $X^{\mathrm{pred}}_{1:n_{\mathrm{w}}}$ to predict an optimized trajectory $\check{X}_{1:n_{\mathrm{p}}}$, while satisfying the potential dynamics constraints and estimating the time allocation of the optimized trajectory.

There are several goals that must be accomplished by the optimization problem. First, trajectory information about the observed mobile robot should be provided, as demonstrated by adequate trajectory waypoints. The number of to be generated trajectory waypoints n_{p} should be specified by the user, which is usually larger than n_{w}, because more waypoints are added between the previous estimated path waypoints to provide a more refined trajectory.

Moreover, the total travel time t_{f} and each trajectory waypoint's arrival time must be estimated during the optimization problem. There are two possible strategies for allocating time. On the one hand, a sequence of time step variables t_p with $p \in [1, n_{\mathrm{p}} - 1]$ can be introduced, and the total travel time can be calculated by the sum of all time variables, implying $t_{\mathrm{f}} = \sum_{1}^{n_{\mathrm{p}}-1} t_p$. On the other hand, the total travel time t_{f} can be set up as an optimization variable, and the time interval Δt between adjacent trajectory waypoints is assumed to be constant, calculated by $\Delta t = \frac{t_{\mathrm{f}}}{(n_{\mathrm{p}}-1)}$. The first strategy is more intuitive, but more difficult to compute by an optimization, and it may violate some numerical approaches resulting in infeasible trajectories [8]. This work is built upon the second strategy and sets up a constant time interval between the adjacent predicted trajectory waypoints.

The observed mobile robot's dynamics and physical limitations must be satisfied. For instance, the predicted trajectory should adhere to the potential dynamics constraints like the maximal absolute velocity and maximal applied forces must be satisfied. Besides, the predicted trajectory should traverse the path waypoints in correct order.

To accomplish the aforementioned goals, this work formulates the optimization problem in a discrete mechanics and optimal control (DMOC) framework with complementarity constraints.

3.1 DMOC Framework

In general, there are different approaches for incorporating the system dynamics of the observed mobile robot into the optimization problem. On the one hand, one can describe the robot's dynamics through the Newton-Euler equations, as demonstrated in our previous work [17] or based on the Euler-Lagrange equations [6]. There, the dynamics is represented by ordinary differential equations. On the other hand, one can employ the Lagrange-d'Alembert principle, then the system dynamics is determined with variational principles [19], which are later considered as equality constraints in the optimization problem. As suggested in [12, 15], the discrete mechanics and optimal control (DMOC) framework can be utilized to plan and estimate an optimized trajectory for mobile robots robustly, which discretizes system dynamics first with the constant time interval Δt using the discrete Lagrange-d'Alembert principle. Thus, we here describe the system dynamics within this framework, and later compare its performance to that of our previous implementation.

In Lagrangian mechanics, a mechanical system can be described using generalized coordinates, which is denoted by a vector $\boldsymbol{q}(t) \in Q$, where Q is the configuration manifold of the system [19]. Then, a motion of this system can be represented by a curve along the manifold, where the system is moved from an initial point $\boldsymbol{q}(t_0)$ to a final point $\boldsymbol{q}(t_f)$ via control forces over the time interval $[0, t_f]$. During this movement, the system is constrained by the Lagrange-d'Alembert principle

$$\delta \int_0^{t_f} L(\boldsymbol{q}(t), \dot{\boldsymbol{q}}(t)) dt + \int_0^{t_f} \boldsymbol{f}(t) \delta \boldsymbol{q}(t) dt = 0, \qquad (9)$$

where L is the Lagrange function (also called Lagrangian in [19]) of the mechanical system that is represented in the form kinetic minus potential energy of the system. The external forces on the system are denoted as $\boldsymbol{f}(t)$.

For the numerical solution with the Lagrange-d'Alembert principle, a discrete formulation of Eq. (9) is specified in the DMOC framework. A sequence of discrete points $\{\boldsymbol{q}_k\}_{k=1}^{n_p}$ is defined that approximates the continuous point $\boldsymbol{q}(t)$, where n_p denotes the number of trajectory points connecting the initial and the final states. Similarly, the applied forces are also discretized and denoted by $\{f_k\}_{k=1}^{n_p}$. Based on the discrete Lagrange-d'Alembert principle, the dynamics constraint in Eq. (9) is converted to the forced discrete Euler-Lagrange equation as in [20]

$$\frac{\partial L_d(q_{k-1}, q_k)}{\partial q_k} + \frac{\partial L_d(q_k, q_{k+1})}{\partial q_k} + f_{k-1}^+ + f_k^- = 0, \text{with } k = [2, n_p - 1], \quad (10)$$

where L_d is the discrete Lagrange function and the external forces are represented at each discrete time step by the left/right forces, denoted by f_k^- and f_k^+, respectively [18]. The initial boundary condition can be deduced using the discrete Legendre transforms [20], resulting in

$$\frac{\partial L(q, \dot{q})}{\partial \dot{q}}\bigg|_{q = q(t_0),\, \dot{q} = \dot{q}(t_0)} + \frac{\partial L_d(q_1, q_2)}{\partial q_1} + f_1^- = 0, \quad (11)$$

given the initial conditions of the system $q(t_0)$ and $\dot{q}(t_0)$.

The discrete Lagrange function and the discrete left/right forces in Eqs. (10) and (11) are specified as [20]

$$L_d(q_k, q_{k+1}) = \Delta t L\left(\bar{q}_k, \dot{\bar{q}}_k\right) \quad \text{and} \quad f_k^- = f_k^+ = \frac{\Delta t}{4}\left(f_{k+1} + f_k\right), \quad (12)$$

where the approximated point and its time derivative are denoted as $\bar{q}_k$ and $\dot{\bar{q}}_k$, which can be estimated based on the midpoint rule

$$\bar{q}_k = \frac{q_k + q_{k+1}}{2} \quad \text{and} \quad \dot{\bar{q}}_k = \frac{q_{k+1} - q_k}{\Delta t}. \quad (13)$$

Finally, the DMOC framework in this work is defined as follow

$$\min_{t_f^*, u^*, q^*} \quad J_d(t_f, u, q)$$

$$\text{subject to} \quad \Delta t = t_f/(n_p - 1),$$
$$q_0 = q(t_0),$$
$$q_{min} \le q_k \le q_{max}, \quad \forall k \in [1, n_p],$$
$$\dot{q}_{min} \le \dot{\bar{q}}_k \le \dot{q}_{max}, \quad \forall k \in [1, n_p - 1], \quad (14)$$
$$u_{min} \le u_k \le u_{max}, \quad \forall k \in [1, n_p],$$
$$\text{discrete mechanics based on Eqs. } (10)-(13),$$
$$\text{and further necessary constraints,}$$

where J_d denotes the objective function of the optimization problem, and the control input of the system at each time step is represented with u_k. Note that the external forces f are dependent on u.

In this work, the proposed algorithm for two typical mobile robots are demonstrated, namely a holonomic omnidirectionally driven mobile robot and a nonholonomic differential wheeled mobile robot, to verify the performance of the proposed method. Their modeling is presented in the following.

Modeling of an Omnidirectional Mobile Robot. An omnidirectional mobile robot can be described as a holonomic mechanical system. The generalized coordinates for the omnidirectional mobile robots can be selected as $\boldsymbol{q} := [x,\ y]^\top$ and its time derivative is denoted as $\dot{\boldsymbol{q}} := [\dot{x},\ \dot{y}]^\top$, when the orientation is ignored. Then, the Lagrange function can be computed by

$$L(\boldsymbol{q}, \dot{\boldsymbol{q}}) = \frac{1}{2} m_{\text{robot}} \dot{\boldsymbol{q}}^\top \dot{\boldsymbol{q}}. \tag{15}$$

Furthermore, the control input for such a system can be defined as $\boldsymbol{u}(t) := [f_{\text{robot}}, \zeta]^\top$, where f_{robot} is the collective force on the mobile robot and the angle ζ is the angle between the collective force and the x-axis of the inertial frame of reference. Then, external forces on the omnidirectional mobile robot can be represented as

$$\boldsymbol{f}(t) = \begin{bmatrix} \cos(\zeta) f_{\text{robot}} \\ \sin(\zeta) f_{\text{robot}} \end{bmatrix}. \tag{16}$$

Modeling of a Nonholonomic Mobile Robot. A differential wheeled mobile robot is a typical nonholonomic mechanical system since the wheels cannot slip sideways. Taking the robot's base and two attached wheels as three individual components, the generalized coordinates of this system are selected as $\boldsymbol{q} := [x,\ y,\ \psi,\ \phi_l,\ \phi_r]^\top$, where ψ denotes the heading of the mobile robot, and the angles of the left and right wheels are denoted as ϕ_l and ϕ_r, respectively. The control input is specified as the generated torques from the onboard DC motors, which is denoted as $\boldsymbol{u}(t) := [\tau_l,\ \tau_r]^\top$.

The Lagrange function of such a system then can be computed by

$$L(\boldsymbol{q}, \dot{\boldsymbol{q}}) = \frac{1}{2} J_{\text{wheel}}(\dot{\phi}_l^2 + \dot{\phi}_r^2) + \frac{1}{2} J_{\text{robot}} \dot{\psi}^2 + \frac{1}{2} m_{\text{robot}}(\dot{x}^2 + \dot{y}^2), \tag{17}$$

where the inertia of the whole mobile robot and an individual wheel are denoted as J_{robot} and J_{wheel}, respectively. Furthermore, for the differential wheeled mobile robot, the kinematic connection between the robot state and the angle velocity of each wheel can be estimated by

$$\begin{bmatrix} \dot{x} \\ \dot{y} \\ \dot{\psi} \end{bmatrix} = \begin{bmatrix} \dfrac{\cos(\psi) r_{\text{wheel}}(\dot{\phi}_l + \dot{\phi}_r)}{2} \\ \dfrac{\sin(\psi) r_{\text{wheel}}(\dot{\phi}_l + \dot{\phi}_r)}{2} \\ \dfrac{r_{\text{wheel}}(-\dot{\phi}_l + \dot{\phi}_r)}{l_{\text{wheel}}} \end{bmatrix}, \tag{18}$$

where the distance between two wheels is represented as l_{wheel} and the radius of the wheel is denoted as r_{wheel}.

Bringing Eq. (18) into Eq. (17), the corresponding reduced coordinates of the system are selected as $\boldsymbol{r} := [\phi_l,\ \phi_r]^\top$. The constrained reduced Lagrange function is defined in [3, 15] as

$$L_c(\boldsymbol{r}, \dot{\boldsymbol{r}}) = \frac{1}{2} \left(J_{\text{wheel}} + \frac{r_{\text{wheel}}^2 m_{\text{robot}}}{4} + \frac{J_{\text{robot}} r_{\text{wheel}}^2}{l_{\text{wheel}}^2} \right) \left(\dot{\phi}_l^2 + \dot{\phi}_r^2 \right) + \\ \left(\frac{r_{\text{wheel}}^2 m_{\text{robot}}}{4} - \frac{J_{\text{robot}} r_{\text{wheel}}^2}{l_{\text{wheel}}^2} \right) \dot{\phi}_l \dot{\phi}_r, \tag{19}$$

while taking Eq. (18) as an additional constraint in the optimization problem. The external forces on the differential wheeled mobile robot is identical to the control input that satisfies $\boldsymbol{f}(t) = \boldsymbol{u}(t)$.

3.2 Complementarity Constraints

Apart from restraining the observed mobile robot's dynamics in the optimization problem, it is also necessary to enforce that the optimized trajectory traverses the desired path waypoints $\boldsymbol{X}_{1:n_w}^{\text{pred}}$ in the correct sequence. Intuitively, one can specify that a particular optimized trajectory waypoint $\breve{\boldsymbol{x}}_p$ muss coincide with an estimated path waypoint $\boldsymbol{x}_j^{\text{pred}}$. However, this may conflict with the potential system dynamics. To address this issue, in this study the arrangement of the predicted trajectory waypoints will be determined automatically with complementarity constraints by solving the proposed optimization problem.

Generally, a complementarity constraint can be defined as

$$f_1(z) f_2(z) = 0, \quad \text{with} \quad f_1(z) \geq 0 \quad \text{and} \quad f_2(z) \geq 0, \tag{20}$$

or marked equivalently with the complementarity operator $\perp$ as

$$0 \leq f_1(z) \perp f_2(z) \geq 0, \tag{21}$$

where f_1 and f_2 are two specified non-negative functions of the state z. Given Eqs. (20) or (21), at least one the functions f_1 and f_2 should be zero. One can investigate the relationship between each pair of the estimated path waypoint $\boldsymbol{x}_w^{\text{pred}}$ and the optimized trajectory waypoint $\breve{\boldsymbol{x}}_p$ through

$$0 \leq f_{\text{progress}}(\boldsymbol{\Lambda}) \perp f_l(\boldsymbol{x}_w^{\text{pred}}, \breve{\boldsymbol{x}}_p) \geq 0, \tag{22}$$

where the function f_l calculates the Euclidean distance between two specified positions in the xy-plane, and the non-negative function f_{progress} denotes the progress changes with a progress variable set $\boldsymbol{\Lambda} := \{\lambda_p^w \in [0,1] | p = [1, \ldots, n_p], w = [1, \ldots, n_w]\}$. Besides, for an arbitrary estimated path waypoint $\boldsymbol{x}_w^{\text{pred}}$, the boundary conditions of the progress variables hold

$$\lambda_{p=1}^w = 1 \quad \text{and} \quad \lambda_{p=n_p}^w = 0. \tag{23}$$

Thereby, based on the complementarity (22) the relationship between $\boldsymbol{x}_w^{\text{pred}}$ and $\breve{\boldsymbol{x}}_p$ can be indicated by the progress change function f_{progress}, because f_{progress} can be positive if and only if $\breve{\boldsymbol{x}}_p$ has reached the given the waypoint

x_w^{pred}; otherwise, it should be zero to satisfy the complementary condition. In other words, the progress of the optimized trajectory given the estimated path can be organized with an appropriate setup of the progress change function f_{progress}.

As inspired by [8], in this work the progress change function f_{progress} is defined as

$$f_{\text{progress}}(\boldsymbol{\Lambda}) = \lambda_p^w - \lambda_{p+1}^w. \tag{24}$$

Given the complementarity (22), the proposed progress change function f_{progress} in Eq. (24) indicates that the progress variable λ_{p+1}^w could be nonidentical with its previous one λ_p^w, if the p-th optimized trajectory waypoint has arrived at the w-th estimated path waypoint x_w^{pred}. Considering the boundary conditions in Eq. (23), at this time, the progress variable λ_p^w could be changed from its initial value to zero. Moreover, to guarantee that the optimized trajectory travels through the predicted path waypoints in the correct order, the condition

$$\lambda_p^w \le \lambda_p^{w+1}, \ \forall w \in [1, n_{\text{w}} - 1] \text{ and } p \in [1, n_{\text{p}}] \tag{25}$$

must be satisfied, which ensures that the optimized $\check{x}_p$ passes the waypoint x_w^{pred} before proceeding to the next waypoint x_{w+1}^{pred}.

Additionally, optimization problems involving complementarity constraints are usually difficult to solve, as the classical constraint qualifications are failed to hold due to the complementarity constraints [13]. In this work, additional relaxation variables ν_p^w are introduced to mitigate the effect arising from the complementarity constraints. Thus, the complementarity (22) is converted to

$$0 \le f_{\text{progress}}(\boldsymbol{\Lambda}) \perp (f_l(x_w^{\text{pred}}, \check{x})_p - \nu_p^w) \ge 0. \tag{26}$$

It is not a prudent strategy to force the optimized trajectory $\check{X}$ to pass through all the estimated path waypoints precisely. On the one hand, the predicted waypoints may contain outliers, affecting the outcome of the optimized trajectory significantly. On the other hand, the accuracy of the predicted waypoints is limited by the grid size from the last stage. Therefore, each relaxation variable ν_p^w must satisfy

$$0 \le \nu_p^w \le d_{\text{tolerance}} \ \forall p \in [1, n_{\text{p}} - 1], \tag{27}$$

in which $d_{\text{tolerance}}$ denotes the maximum acceptable deviation from the estimated path waypoint.

In Fig. 6, for instance, an ideal distribution of the progress variables is depicted, where three estimated path waypoints x_w^{pred} are denoted by green squares and the optimized trajectory waypoints $\check{x}_p$ are signified by black circles. The progress variables λ_p^2 for the estimated path waypoint x_2^{pred} can be reduced to zero up the fourth component λ_4^2 since the trajectory waypoint $\check{x}_3$ becomes sufficiently close to its counterpart. Similarly, the progress variables λ_p^3 for the waypoint x_3^{pred} can vary only at the last one according to the constraint definitions in (22) and (23). Furthermore, all results comply with the constraint inequality (25), ensuring that the produced optimal trajectory passes through the x_2^{pred} first.

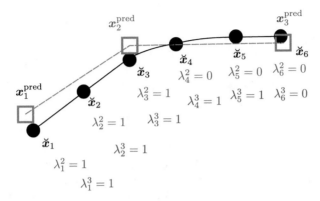

Fig. 6. The progress variables associated with the estimated path waypoint x_2^{pred} and x_3^{pred} that are marked red and blue, respectively.

3.3 Optimization Formulation

In this work, the optimization problem is formulated using the discrete mechanics and optimal control (DMOC) framework in Eq. (14). The observed mobile robot is described with the vector q in the individual generalized coordinates, where the first two elements of q represent the optimized position of the mobile robot $(\breve{x}, \breve{y})$ in the xy-plane. Besides, the optimization parameters X^{opt} include the total travel time t_{f}, the progress variable λ_p^w, and the relaxation variable ν_p^w at each discrete time step.

Compared to our previous implementation in [17], which included the sum of all progress variables, the cost function of the optimization problem in this work contains the total travel time t_{f} and the total travel distances. To prevent the non-ideal predicted waypoint distribution shown in [17], the initial value of all optimization parameters X^{opt} are estimated beforehand based on the linear interpolation, which also significantly reduce the number of iterations to find the optimal solution. Based on the introduction in Sects. 3.1 and 3.2, the optimization problem is formulated as

$$\min_{X^{\text{opt}}} \quad \gamma_1 t_{\text{f}} + \gamma_2 \sum_{l=1}^{n_{\text{p}}-1} (\|\breve{x}_{l+1} - \breve{x}_l, \breve{y}_{l+1} - \breve{y}_l\|_2^2)$$

$$\text{subject to} \quad \text{constraints defined in Eq. (14),}$$

$$\text{and further constraints based on Eqs. (23)–(27),}$$

(28)

where the parameters $\gamma_{1/2}$ specify the weights for the total travel time and the trajectory length, respectively. Notably, this implementation does not imply providing optimal planning for the observed mobile robot, since as the invested scenario, the observer can neither control its observing target nor affect its motion but rather finds a sequence of possible trajectory waypoints that passes through the estimated path waypoint from the last stage in an effective manner while complying with the potential dynamics.

To solve the proposed optimization problem, CasADi [1] is utilized in conjunction with the solver IPOPT [25]. DMOC has demonstrated its advantages for solving a long horizon optimization problem robustly to determine the optimal results in the following example, see Fig. 7. The proposed optimization formulation and the one from our previous work [17] are utilized to estimate a sequence of the optimized trajectory for the omnidirectional mobile robot, given five path waypoints and the same initial configurations, see Fig. 7(a). In Fig. 7(b), the estimated total travel time t_f is shown with different numbers of trajectory waypoints n_p. The proposed method has more stable prediction results; meanwhile, the IPOPT solver can always find the optimal solution in this scenario. On the contrary, the formulation in [17] may encounter some numerical issues, and thus the predicted results there are not as stable as in our novel implementation.

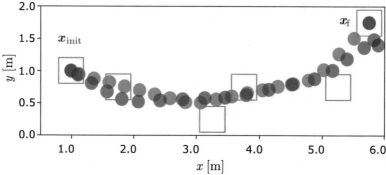

(a) Predicted trajectory of the proposed optimization formulation (●), and the predicted results based on [17] (●). Moreover, the estimated path waypoints from the last stage are denoted with □, and the number of trajectory waypoints n_p for both implementations is set to be 20.

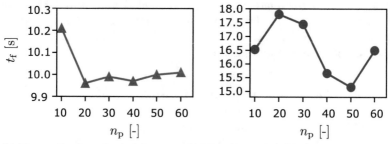

(b) The predicted total travel time t_f with different number of trajectory waypoints n_p given several path waypoints from the last stage. The results based on the proposed optimization framework of this work and our previous work are denoted with ▲ and ●, respectively.

Fig. 7. Performance comparison between the proposed optimization formulation and our previous implementation in [17].

4   Experiments

Several experiments are conducted to validate the proposed two-stage approach in this work, including simulation experiments in the Gazebo environment and real-world experiments with our self-designed mobile robots as illustrated in Fig. 8. The omnidirectional mobile robot used in the experiment is our **H**olonomic **E**xtensible **R**obotic **A**gent (HERA), driven omnidirectionally through four Mecanum wheels attached at 90° from each other on the base layer [7]. The differential wheeled mobile robot is our **DI**fferentially **A**ctuated **N**onholonomic **A**gent (DIANA), which is driven by two DC motors and two wheels fixed to the robot's base [21]. Throughout the hardware experiments, both robots are controlled with onboard computing unit Beaglebone Blue, and their current position and orientation are measured using the OptiTrack indoor localization system. In the simulation experiment, both robots are simulated in the Gazebo environment with a physical engine using their 3D geometrical and mechanical models.

(a) Customized mobile robots   (b) Simulated mobile robots

Fig. 8. Robots in the experiment [7,21].

4.1   Simulation Experiments

For the simulation, a large scenario is built within Gazebo environment that has the size of 20 m × 6 m, see Fig. 9(a). In the scenario, several obstacles are placed, and four probable destinations are provided that are marked with red stars in Fig. 9(b). Throughout the experiment, both HERA and DIANA robots are controlled manually with a joystick from the initial position to the destination at the top right corner, and the recorded trajectories of the two robots are illustrated in Fig. 9(b) with different markers. Due to the individual dynamics features of the HERA and the DIANA robots and the manual steering, the trajectories show differences, most notably when turning around obstacles.

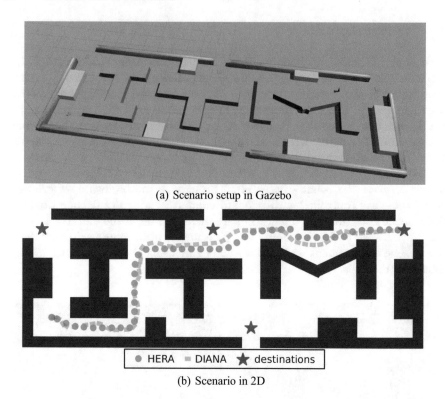

(a) Scenario setup in Gazebo

● HERA   ■ DIANA   ★ destinations

(b) Scenario in 2D

Fig. 9. Simulation scenario in Gazebo.

Keep in mind, that not the ground robots are controlled automatically but that their intention should be guessed and what the ground robots intend to do.

In the first stage, since the scenario is identical for both mobile robots, the sampling procedure is generally equal for both HERA and DIANA robots, although the initial position of each sampling may differ. Thus, in Fig. 10(a), just the sampling results from the HERA robot are given, where the potential destinations are denoted with olive squares, while the gradation of color indicates the estimated intention of the observed robot to each destination. In Figs. 10(b) and 10(c), the estimated path waypoints according to the proposed procedure in Sect. 2 for each robot are shown. The gray squares denote the most frequently visited grid at each step from the robot's current position to the destination with the highest probability; meanwhile, the path waypoints based on greedy strategy are shown with circles. Compared to the greedy results, the sum-pooling results avoid too close connections that might make subsequent optimization in the second stage difficult to solve. In general, the estimated path waypoints for both robots are relatively similar, as only the potential geometrical connections are determined based on the proposed Bayesian framework on the discrete roadmap in the first stage.

After estimating the potential path reference from the first stage, the proposed method in the second stage is applied, and by solving the proposed optimization problem, the predicted trajectories are illustrated in Fig. 11. Although the estimated path waypoints do not reveal significant differences in the dynamics of the two robots, the predicted trajectory for each robot does differ, in particular, when turning around obstacles. The predicted result for the omnidirectional mobile robot is more straightforward, passing through each estimated path waypoints within the tolerance; on the contrary, limited by the differential wheeled robot's kinematic features, the predicted results for the DIANA robot demonstrate that the robot needs to make a smoother turn to change its direction when turning around obstacles, which is consistent with the actual simulation trajectory.

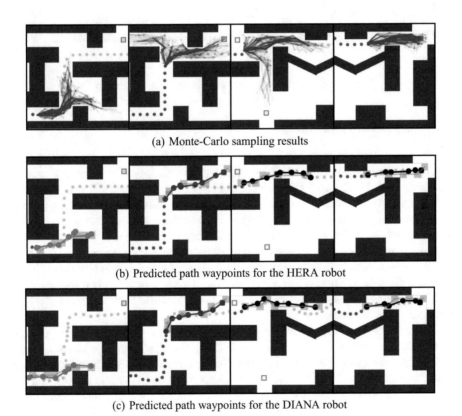

(a) Monte-Carlo sampling results

(b) Predicted path waypoints for the HERA robot

(c) Predicted path waypoints for the DIANA robot

Fig. 10. Simulation experimental results in stage one.

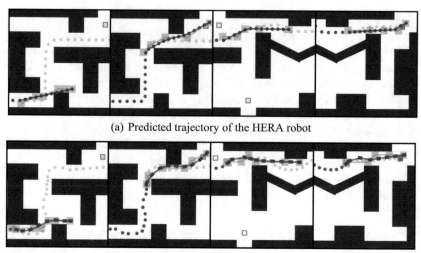

(a) Predicted trajectory of the HERA robot

(b) Predicted trajectory of the DIANA robot

Fig. 11. Simulation experimental results in stage two.

4.2   Hardware Experiment

In the hardware experiment, a scenario is set up within a $3\,\text{m} \times 3\,\text{m}$ area with a T-shaped obstacle, see Fig. 12(a). Similar to the simulation in the Gazebo environment, both HERA and DIANA robots are controlled remotely and manually through a joystick, and there are two alternative destinations (θ_1 and θ_2) in the scene denoted by red stars in Fig. 12(b). The recorded trajectories of each robot are denoted with the individual markers, which are also due to the different dynamics properties. Due to the HERA robot's ability to travel in any direction flexibly, which makes it more suitable for use in this type of narrow scenario, its path is significantly shorter than that of the DIANA robot. On the contrary, the DIANA robot requires a large radius of rotation in order to navigate around the obstacle in the scenario.

After processing the proposed algorithm, the predicted trajectories of each robot are illustrated in Figs. 12(c) and 12(d). In this scenario, there is more open space than in the simulation scenario; as a result, the predicted path waypoints are more direct to the destination and more sensitive to the distribution of the nodes of the roadmap. At the very beginning, since the destination θ_1 is much closer to the initial position of the robot, our proposed algorithm determines wrongly that it might be the most likely destination, and the predicted trajectory is heading to θ_1. After passing by the destination θ_1, the proposed algorithm can sensitively notice the destination's change of each destination and make a reasonable prediction to θ_2 in the future in comparison to the ground truth trajectory. Regarding the performance in simulation, for the HERA robot, our proposed algorithm generates the predicted trajectory more straightforward

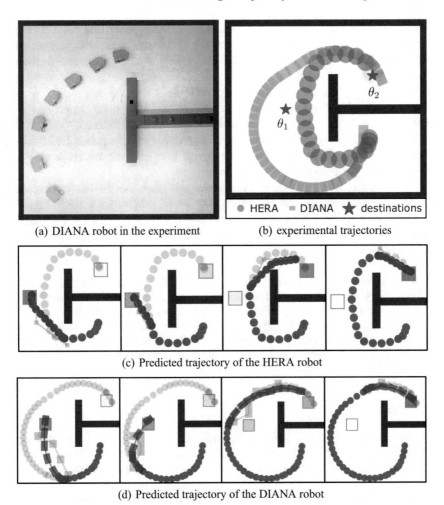

(a) DIANA robot in the experiment   (b) experimental trajectories

(c) Predicted trajectory of the HERA robot

(d) Predicted trajectory of the DIANA robot

Fig. 12. Hardware experiment.

through estimated path waypoints since it can move any direction with its Mecanum wheels. On the contrary, the prediction for the DIANA robot adheres to its dynamic limitations while adjusting the heading to the destination within the tolerance.

5   Conclusion

In this work, we offer a two-stage algorithm for predicting the future trajectories of an observed mobile robot based on its previous trajectories. The proposed algorithm determines the potential destination given the scenario information and the observed robot's previous paths and then uses the Monte-Carlo sampling process to estimate a guessed future path. Then, in the second stage, an optimization problem is formulated based on the DMOC framework, which determines a more reasonable smoother predicted trajectory according to the individual dynamics properties and the previously estimated path. Furthermore, compared to our previous work, the proposed approach can solve the optimization problem more robustly. Finally, in both simulation and hardware experiments, we investigate two typical mobile robots, the holonomic omnidirectionally driven mobile robot HERA and the nonholonomic wheeled mobile robot DIANA, and demonstrate that our proposed algorithm can predict their future trajectories consistent with ground-truth trajectories.

In the future, one may also incorporate the obstacles in the optimization formulation in the second stage, which may avoid predicting a trajectory that is too close to obstacles. Moreover, the proposed algorithm can be implemented in three-dimensional space to predict the trajectory of aerial devices, such as quadrotors and airplanes.

Acknowledgements. This work was supported by the Deutsche Forschungsgemeinschaft (DFG, German Research Foundation) under Grant 433183605 and through Germany's Excellence Strategy (Project PN4-4 Theoretical Guarantees for Predictive Control in Adaptive Multi-Agent Scenarios) under Grant EXC 2075-390740016. This research also benefited from the support by the China Scholarship Council (CSC, No. 201808080061) for Wei Luo. We also would like to thank Dr. Henrik Ebel and Mr. Mario Rosenfelder for supporting the HERA and DIANA mobile robots.

References

1. Andersson, J.A.E., Gillis, J., Horn, G., Rawlings, J.B., Diehl, M.: CasADi - a software framework for nonlinear optimization and optimal control. Math. Program. Comput. **11**(1), 1–36 (2019). https://doi.org/10.1007/s12532-018-0139-4
2. Best, G., Fitch, R.: Bayesian intention inference for trajectory prediction with an unknown goal destination. In: IEEE/RSJ International Conference on Intelligent Robots and Systems (IROS), Hamburg, Germany, pp. 5817–5823, September 2015
3. Bloch, A.M.: Nonholonomic Mechanics and Control. Springer, New York (2015). https://doi.org/10.1007/978-1-4939-3017-3
4. Dai, S., Li, L., Li, Z.: Modeling vehicle interactions via modified LSTM models for trajectory prediction. IEEE Access **7**, 38287–38296 (2019)
5. Dendorfer, P., Ošep, A., Leal-Taixé, L.: Goal-GAN: multimodal trajectory prediction based on goal position estimation. In: Asian Conference on Computer Vision (ACCV). Virtual Conference, pp. 1–17, November 2020
6. Dhaouadi, R., Hatab, A.A.: Dynamic modelling of differential-drive mobile robots using Lagrange and Newton-Euler methodologies: a unified framework. Adv. Robot. Autom. **02**(02), 1–7 (2013)
7. Ebel, H., Luo, W., Yu, F., Tang, Q., Eberhard, P.: Design and experimental validation of a distributed cooperative transportation scheme. IEEE Trans. Autom. Sci. Eng. **18**(3), 1157–1169 (2021)
8. Foehn, P., Scaramuzza, D.: CPC: complementary progress constraints for time-optimal quadrotor trajectories. In: Robotics: Science and Systems. Virtual Conference, pp. 1–12, July 2020

9. Giuliari, F., Hasan, I., Cristani, M., Galasso, F.: Transformer networks for trajectory forecasting, pp. 1–8 (2020). arXiv:2003.08111
10. Gombo, Y., Tiwari, A., Devasia, S.: Accelerated-gradient-based flexible-object transport with decentralized robot teams. IEEE Robot. Autom. Lett. **6**(1), 151–158 (2021)
11. Gupta, A., Johnson, J., Li, F.F., Savarese, S., Alahi, A.: Social GAN: socially acceptable trajectories with generative adversarial networks. In: IEEE Conference on Computer Vision and Pattern Recognition (CVPR), Salt Lake City, USA, pp. 2255–2264, June 2018
12. Junge, O., Marsden, J.E., Ober-Blöbaum, S.: Discrete mechanics and optimal control. IFAC Proc. Vol. **38**(1), 538–543 (2005)
13. Kanzow, C., Schwartz, A.: A new regularization method for mathematical programs with complementarity constraints with strong convergence properties. SIAM J. Optim. **23**(2), 770–798 (2013)
14. Karaman, S., Frazzoli, E.: Sampling-based algorithms for optimal motion planning. Int. J. Robot. Res. **30**(7), 846–894 (2011)
15. Kobilarov, M., Sukhatme, G.: Optimal control using nonholonomic integrators. In: IEEE International Conference on Robotics and Automation (ICRA), Rome, Italy, pp. 1832–1837, May 2007
16. Liu, C., Hedrick, J.K.: Model predictive control-based target search and tracking using autonomous mobile robot with limited sensing domain. In: American Control Conference (ACC), Seattle, USA, pp. 2937–2942, May 2017
17. Luo, W., Eberhard, P.: Optimization-based trajectory prediction enhanced with goal evaluation for omnidirectional mobile robots. In: Proceedings of the 18th International Conference on Informatics in Control, Automation and Robotics. Virtual Conference, pp. 263–273. SCITEPRESS - Science and Technology Publications, July 2021
18. Marsden, J.E., West, M.: Discrete mechanics and variational integrators. Acta Numer **10**, 357–514 (2001)
19. McLachlan, R., Perlmutter, M.: Integrators for nonholonomic mechanical systems. J. Nonlinear Sci. **16**(4), 283–328 (2006). https://doi.org/10.1007/s00332-005-0698-1
20. Ober-Blöbaum, S., Junge, O., Marsden, J.E.: Discrete mechanics and optimal control: an analysis. ESAIM Control Optim. Calc. Var. **17**(2), 322–352 (2011)
21. Rosenfelder, M., Ebel, H., Eberhard, P.: Cooperative distributed model predictive formation control of non-holonomic robotic agents. In: International Symposium on Multi-Robot and Multi-Agent Systems (MRS), Cambridge, UK, pp. 11–19, November 2021
22. Schöller, C., Aravantinos, V., Lay, F., Knoll, A.: What the constant velocity model can teach us about pedestrian motion prediction. Robot. Autom. Lett. (RA-L) **5**(2), 1696–1703 (2020)
23. Su, Z., Wang, C., Cui, H., Djuric, N., Vallespi-Gonzalez, C., Bradley, D.: Temporally-continuous probabilistic prediction using polynomial trajectory parameterization, pp. 1–7 (2020). arXiv:2011.00399
24. Verjans, M., Phlippen, L., Schleer, P., Radermacher, K.: SEBARES - design and evaluation of a controller for a novel externally guided self-balancing patient rescue aid. In: 18th European Control Conference (ECC), Naples, Italy, pp. 209–214, June 2019

25. Wächter, A., Biegler, L.T.: On the implementation of an interior-point filter line-search algorithm for large-scale nonlinear programming. Math. Program. **106**(1), 25–57 (2005). https://doi.org/10.1007/s10107-004-0559-y
26. Yu, C., Ma, X., Ren, J., Zhao, H., Yi, S.: Spatio-temporal graph transformer networks for pedestrian trajectory prediction. In: Vedaldi, A., Bischof, H., Brox, T., Frahm, J.-M. (eds.) ECCV 2020. LNCS, vol. 12357, pp. 507–523. Springer, Cham (2020). https://doi.org/10.1007/978-3-030-58610-2_30

Adaptive Neural Network Based Fractional Order Control of Unmanned Aerial Vehicle

Heera Lal Maurya[✉][ORCID], Padmini Singh[ORCID], Subhash Chand Yogi[ORCID],
Laxmidhar Behera[ORCID], and Nishchal K. Verma[ORCID]

Department of Electrical Engineering, Indian Institute of Technology, Kanpur, Kanpur 208016,
Uttar Pradesh, India
{hlmaurya,padminis,subyogi,lbehera,nishchal}@iitk.ac.in

Abstract. An unmanned Aerial Vehicle (UAV) is a highly non-linear unstable
system. In this work using fractional order calculus, a novel fractional order
dynamics of UAV is proposed. The concept of fractional order depicts the more
realistic behavior of UAVs. For proposed fractional order model, a fractional
order sliding mode controller (SMC) is designed such that the desired path can be
achieved by the UAV in finite-time. In addition to this an adaptive neural network
(ANN) based approximation function is attached to the controller having the quality of optimal hidden nodes. The weight associated to the hidden nodes achieves
the optimal values. The integration of ANN based function with fractional order
SMC achieves better results compared to fractional order SMC alone. Stability
analysis is given for the fractional order SMC using fractional Lyapunov method.
The same Lyapunov function has been used for finding the adaptive law for estimating the unknown dynamics of the system. Simulations have been done for
position and attitude tracking of UAV using ANN based fractional order SMC to
demonstrate the advantage of the proposed method.

Keywords: Unmanned aerial vehicle · Adaptive neural network · Fractional
order control · Sliding mode control

1 Introduction

Recently, UAVs are being used for wide variety of applications some of them are transportation, surveillance [1], forest trail detection [6], agriculture purposes [13] etc. however to control a quadrotor is quite challenging due to its characteristics like high nonlinearity, underactuation property and external disturbances. From the past few years it
is a subject of interest for researchers to design a robust controller for quadrotor UAVs.

Although there are several controllers e.g. LQR controller [5], Backstepping Controller [11,24] developed and applied on the UAV, still a robust control scheme has been
an interest of research. Sliding mode control (SMC) [16] is one of the most popular and
robust control technique which has the ability to rejects disturbances and uncertainties
but at the cost of chattering [2]. The chattering actuates the unwanted dynamics of the
system which can deteriorate the system performance, hence disturbance rejection at a
cost of deteriorated performance is not appreciable. Since quadrotor is a relative degree

O. Gusikhin et al. (Eds.): ICINCO 2021, LNEE 1006, pp. 63–79, 2023.
https://doi.org/10.1007/978-3-031-26474-0_4

two type of system, a proper stable sliding surface is needed for the design of controller. Depending upon the type of surface the convergence of error can be asymptotic or finite-time. The asymptotic surface [23] shows slower convergence than the finite-time surface but finite-time surface or terminal sliding surface [21,22] cause singularity issue.

Most of the controllers discussed above are integer order control schemes. Recently, fractional order controllers [3,4,8] have drawn much attention due to application of powerful processors. The fractional order terms provides an extra degree of freedom in terms of controller parameters which can be adjusted for better tracking performance. Some of the work on fractional order controller on UAV are as [15] presents a PI fractional order controller for quadrotor for only attitude control. A novel fractional controller has been proposed in [9] for attitude control as well as position control of quadrotor. Conventional SMC has been employed in [17] for position and attitude control of quadrotor where fractional order switching law is proposed to compensate the uncertainties on integer model of quadrotor. A fast terminal SMC (FTSMC) has been presented in [10] for faster convergence of tracking error however FTSMC is applied only on attitude control whereas conventional SMC is employed for position control and thus overall scheme doesn't provide faster convergence. All these schemes have been applied on applied on the integer order model of the quadrotor UAV. Moreover, less attention is given to the fractional-order based dynamics. Though we consider the dynamics of the quadrotor of integer order but practically it may not be of integer order because there may be some fractional order term exist which effects the dynamics of the quadrotor. Hence, a fractional order controller can increase the robustness and usability of controller if employed with the fractional order model of the quadrotor.

Motivated from the above discussion, a fractional order model of quadrotor has been considered instead of an integer order model In this work which is more practical and feasible with the real world model. Thereafter, A robust control law as fractional order sliding mode controller (SMC) has been presented for the quadrotor model while considering the uncertain dynamics. There is a trade-off has been done between the asymptotic surface and finite-time time surface using fractional order theory. Using fractional order theory, a novel fractional sliding surface is proposed for the quadrotor which improves the response of the surface as well as avoids the singularity issue of the finite-time sliding surface. Next for mitigating the chattering issue of SMC, a power rate reaching law along with a proportional term has been used in the control laws for position and attitude tracking. There are six fractional order control laws are designed for UAV where, three are position controllers which generates the thrust required and attitude reference for attitude controller while rest of the three are attitude controllers. This work is presented in the paper [12]. In this work extended version of the paper [12] is presented.

Adaptive neural network architecture is used for estimating the parameters of the unknown system or for predicting the dynamics of the system [7]. Adaptive control law is derived using Lyapunov function. Compared to ANN based integer order SMC ANN based fractional order SMC gives more accurate result. In the proposed architecture optimal hidden nodes are used to optimize the weight updation process of the neural network and adaptive feed forward neural network architecture is used for estimating the parameters [20].

The main contribution of the paper are summarised as follows:

1. A fractional order novel sliding surface is proposed so that the region of stability increased in left half plane.
2. A novel fractional order singularity free control law is proposed which rejects the model uncertainties present in the quadrotor.
3. Stability of the fractional sliding surface and the controller is given using Lyapunov stability theory.
4. Optimal number of hidden nodes are used for the adaptive feedforward neural network to make the system adaptive in nature. Adaptive tuning laws are estimated using Lyapunov stability criterion.
5. Simulations are conducted for quadrotor position and attitude tracking and it is shown that the proposed technique is better than the existing fractional order SMC.

Rest of the paper is organized as follows: preliminaries for fractional calculus is provided in Sect. 2 and Sect. 3 constitutes the problem formulation. Fractional order quadrotor model is presented in Sect. 4 which is followed by the controller design in Sect. 5 and stability analysis in Sect. 6. Neural network based adaptive controller is given in Sect. 7. Simulation results along with the comparative analysis has been presented in Sect. 8. Finally, conclusion and future scope is given in Sect. 9.

2 Preliminaries

The caputo fractional derivative [14] of any function $f(\vartheta)$ is represented by:

$$D^{\beta}[f(\vartheta)] = \frac{1}{\Gamma(n-\beta)} \int_{t_0}^{t} \frac{f^{(n)}(\tau)}{(t-\tau)^{(\beta-n+1)}} d\tau \tag{1}$$

where, $\Gamma(.)$ represents the Gamma function and $n-1 < \beta < n \in N$.

Fractional order integral [14] of order $\alpha > 0$ can be expressed as:

$$J^{\alpha}[f(\vartheta)] = \frac{1}{\Gamma(\alpha)} \int_{t_0}^{t} (t-\tau)^{(\alpha-1)} f(\vartheta) d\tau \tag{2}$$

3 Problem Formulation

The control architecture is shown in Fig. 2 consists of two control loops where first one corresponds to inner or attitude control loop which runs at high frequency, another is position control loop which estimates the attitude reference to the attitude controller in terms of ϕ_d, θ_d and thrust to the quadrotor. The speed of the rotor is regulated by the pulse width modulated (PWM) signal. The generation of the desired PWM signal is controlled by the output of attitude controller i.e. torque and thrust from the position controller, which then actuates the motors of quadrotor (Fig. 1).

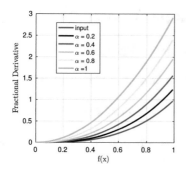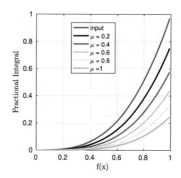

Fig. 1. [19] fractional derivative and fractional integral plot of $f(x) = x^3$.

The objective is to track the desired position and attitude in presence of uncertainties where the desired position x_d, y_d, z_d and desired yaw angle ψ_d are provided by the user. Here a novel fractional order sliding surface is proposed which increases the stability range of the error plane. After that a fractional order adaptive neural network based SMC is applied on the quadrotor which rejects the uncertainty present in the system.

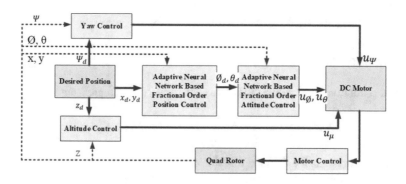

Fig. 2. Control architecture for UAV.

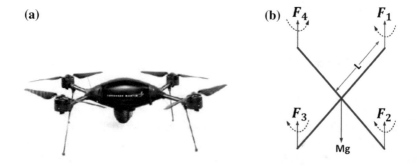

Fig. 3. [12] (a) Quadrotor (b) Direction of forces acting on four arms.

4   Quadrotor Model

Figure 3 represents the pictorial view of quadrotors and direction of the forces acting on the four arms. The forces F_1, F_2, F_3 and F_4 works in the upwards direction to generate the desired thrust so that the quadrotor can fly in the upwards direction. The sum of the total forces $u_\mu = F_1 + F_2 + F_3 + F_4$ is called the total thrust required to lift the quadrotor. The minimum thrust required to drag the UAV in the upward direction should be greater than the weight of the UAV. Therefore small UAVs required small thrust compared to big one to achieve the same height from the ground and hence takes less power.

4.1   Quadrotor Dynamics

Generally fractional order controllers are designed for integer order system. In [8] fractional order sliding mode controller is designed for integer order UAVs. It will be more realistic if one would take the dynamics of UAVs also fractional order. The fractional order dynamics of quadrotor in presence of model uncertainty and external disturbance, is presented here by taking the fraction order of the model of UAV given in [18]. It is to be noted that in this paper a novel fractional model of the quadrotor is designed.

$$D^\alpha \phi_1 = \phi_2$$

$$D^\alpha \phi_2 = \delta f(\phi, t) + d_\phi(t) + \dot\theta\dot\psi(\frac{J_y - J_z}{J_x}) + \frac{u_\phi}{J_x} \tag{3}$$

$$D^\alpha \theta_1 = \theta_2 \tag{4}$$

$$D^\alpha \phi_1 = \phi_2$$

$$D^\alpha \phi_2 = \delta f(\phi, t) + d_\phi(t) + \dot\theta\dot\psi(\frac{J_y - J_z}{J_x}) + \frac{u_\phi}{J_x}$$

$$D^\alpha \theta_1 = \theta_2$$

$$D^\alpha \theta_2 = \delta f(\theta, t) + d_\theta(t) + \dot\phi\dot\psi(\frac{J_z - J_x}{J_y}) + \frac{u_\theta}{J_y}$$

$$D^\alpha \psi_1 = \psi_2$$

$$D^\alpha \psi_2 = \delta f(\psi, t) + d_\psi(t) + \dot\phi\dot\theta(\frac{J_x - J_y}{J_z}) + \frac{u_\psi}{J_z} \tag{5}$$

$$D^\alpha z_1 = z_2$$

$$D^\alpha z_2 = \delta f(z, t) + d_z(t) + \frac{u_\mu}{m}(C\phi C\theta) - g$$

$$D^\alpha x_1 = x_2$$

$$D^\alpha x_2 = \delta f(x, t) + d_x(t) + \frac{u_\mu(C\phi S\theta C\psi + S\phi S\psi)}{m}$$

$$D^\alpha y_1 = y_2$$

$$D^\alpha y_2 = \delta f(y, t) + d_y(t) + \frac{u_\mu}{m}(C\phi S\theta S\psi - S\phi C\psi)$$

where $C(.), S(.)$ corresponds to $cos(.)$ and $sin(.)$ respectively, ϕ_1, θ_1, ψ_1 are three attitude angles i.e. roll, pitch and yaw angles respectively whereas x_1, y_1, z_1 are positions of the quadrotors. There are total twelve states including angular velocities ϕ_2, θ_2, ψ_2 and translational velocities x_2, y_2, z_2. All these twelve states are controlled by four control inputs u_μ, u_ϕ, u_θ and u_ψ. $\delta f(\phi,t), \delta f(\theta,t), \delta f(\psi,t), \delta f(x,t), \delta f(y,t),$ $\delta f(z,t)$ are model uncertainty and $d_\phi(t), d_\theta(t), d_\psi(t)$ are external disturbances. The relation between rotor forces and four control inputs is given in [18]. The dynamics of the quadrotor can be represented as second order fractional subsystems if the virtual control laws are selected as:

$$u_{x1} = \frac{u_\mu}{m}(\cos\phi\sin\theta\cos\psi + \sin\phi\sin\psi)$$

$$u_{y1} = \frac{u_\mu}{m}(\cos\phi\sin\theta\sin\psi - \sin\phi\cos\psi) \tag{6}$$

$$u_{z1} = \frac{u_\mu}{m}(\cos\phi\cos\theta) - g$$

Therefore,

$$u_\mu = m\sqrt{(u_{x1})^2 + (u_{y1})^2 + (u_{z1} + g)^2} \tag{7}$$

After considering all the virtual control inputs the dynamics of the quadrotor can be decoupled using six second order subsystems. Now objective is to design tracking controller such that desired positions and attitude angles are achieved.

4.2   Error Model

In this section error dynamics of x, y, z and ϕ, θ, ψ is given. Now, the second order fractional dynamics for x position is:

$$D^\alpha x_1 = x_2$$

$$D^\alpha x_2 = \delta f(x,t) + d_x(t) + u_{x1} \tag{8}$$

If the desired x position is x_d then the error will be:

$$e_{x1} = x_1 - x_d$$

$$e_{x2} = x_2 - D^\alpha x_d \tag{9}$$

Hence, fractional order error dynamics for x position in presence of model uncertainty and disturbance is:

$$D^\alpha e_{x1} = e_{x2}$$

$$D^\alpha e_{x2} = \delta f(x,t) + d_x(t) + u_{x1} - D^{2\alpha}x_d \tag{10}$$

Like wise error dynamics for rest of the five subsystems are:

$$D^\alpha e_{y1} = e_{y2}$$

$$D^\alpha e_{y2} = \delta f(y,t) + d_y(t) + u_{y1} - D^{2\alpha}y_d \tag{11}$$

$$D^\alpha e_{z1} = e_{z2}$$
$$D^\alpha e_{z2} = \delta f(z,t) + d_z(t) + u_{z1} - D^{2\alpha} z_d \tag{12}$$

$$D^\alpha e_{\phi1} = e_{\phi2}$$
$$D^\alpha e_{\phi2} = \dot\theta\dot\psi(\frac{J_y - J_z}{J_x}) + \delta f(\phi,t) + d_\phi(t) + u_\phi - D^{2\alpha}\phi_d \tag{13}$$

$$D^\alpha e_{\theta1} = e_{\theta2}$$
$$D^\alpha e_{\theta2} = \dot\phi\dot\psi(\frac{J_z - J_x}{J_y}) + \delta f(\theta,t) + d_\theta(t) + u_\theta - D^{2\alpha}\theta_d \tag{14}$$

$$D^\alpha e_{\psi1} = e_{\psi2}$$
$$D^\alpha_{\psi2} = \dot\phi\dot\theta(\frac{J_x - J_y}{J_z}) + \delta f(\psi,t) + d_\psi(t) + u_\psi - D^{2\alpha}\psi_d \tag{15}$$

Now, in next section fractional order controller is designed for the quadrotor.

5 Controller Design

In this section, a novel robust fractional order SMC has been proposed to counteract parametric uncertainty, external disturbances as well as unmatched uncertainty.

5.1 Fractional Order Sliding Surface Design

The proposed fractional order sliding surface is:

$$s_x(t) = D^{\alpha-1} e_{x2} + D^{\alpha-2} \left[\left[k_{x1}(\mid e_{x1} \mid + \mid e_{x1}) \mid^\beta) + k_{x2} \right. \right.$$
$$\left. (\mid e_{x2} \mid + \mid e_{x2}) \mid^\beta) + (sign(e_{x1})D^{1-\alpha} e_{x2}) \right] sign(e_{x2}) \right] \tag{16}$$

where, $\beta \in (0,1)$ is a positive constant and k_{x1} and k_{x2} are positive tuning parameters. Taking the derivative of sliding surface Eq. (16), we get

$$\dot{s}_x(t) = D^\alpha e_{x2} + D^{\alpha-1} \left[\left[k_{x1}(\mid e_{x1} \mid + \mid e_{x1}) \mid^\beta) + k_{x2} \right. \right.$$
$$\left. (\mid e_{x2} \mid + \mid e_{x2}) \mid^\beta) + (sign(e_{x1})D^{1-\alpha} e_{x2}) \right] sign(e_{x2}) \right] \tag{17}$$

After the reaching phase is achieved i.e. when $\dot{s}_x(t) = 0$, Eq. (17) reduces to,

$$D^\alpha e_{x2} = -D^{\alpha-1} \left[\left[k_{x1}(\mid e_{x1} \mid + \mid e_{x1}) \mid^\beta) + k_{x2}. \right. \right.$$
$$\left. (\mid e_{x2}) \mid + \mid e_{x2}) \mid^\beta) + (sign(e_{x1})D^{1-\alpha} e_{x2}) \right] sign(e_{x2}) \right] \tag{18}$$

Error Dynamics in Sliding Mode. From Eq. (18) and (9), fractional dynamics for x position in sliding mode can be written as:

$$D^\alpha e_{x1} = e_{x2}$$

$$D^\alpha e_{x2} = -D^{\alpha-1}\left[\left[k_{x1}(|\,e_{x1}\,| + |\,e_{x1}\,|^\beta) + k_{x2}.\right.\right.$$

$$\left.\left.(|\,e_{x2}\,| + |\,e_{x2}\,|^\beta) + (sign(e_{x1})D^{1-\alpha}e_{x2})\right]sign(e_{x2})\right]$$

(19)

Likewise, error dynamics for rest of the quadrotor states can be obtained.

5.2   Fractional Order Controller Design

Again revisiting Eq. (17)

$$\dot{s}_x(t) = D^\alpha e_{x2} + D^{\alpha-1}\left[\left[k_{x1}(|\,e_{x1}\,| + |\,e_{x1}\,|^\beta) + k_{x2}.\right.\right.$$

$$\left.\left.(|\,e_{x2}\,| + |\,e_{x2}\,|^\beta) + (sign(e_{x1})D^{1-\alpha}e_{x2})\right]sign(e_{x2})\right]$$

(20)

Substituting value of $D^\alpha e_{x2}$ from Eq. (10) in Eq. (20) and substituting $\dot{s}_x(t) = -k_{x3}s_x - k_{x4}\,|\,s_x\,|^\gamma\,sign(s_x)$ the control law u_{x1} will be:

$$-k_{x3}s_x - k_{x4}\,|\,s_x\,|^\gamma\,sign(s_x) = \delta f(x,t) + d_x(t) + u_{x1}$$

$$-D^{2\alpha}x_d + D^{\alpha-1}\left[\left[k_{x1}(|\,e_{x1}\,| + |\,e_{x1}\,|^\beta) + k_{x2}.\right.\right.$$

$$\left.\left.(|\,e_{x2}\,| + |\,e_{x2}\,|^\beta) + (sign(e_{x1})D^{1-\alpha}e_{x2})\right]sign(e_{x2})\right]$$

(21)

where, $\gamma \in (0,1)$ is a positive constant and k_{x3} and k_{x4} are positive tuning parameters. Further simplifying

$$u_{x1} = -k_{x3}s_x - k_{x4}\,|\,s_x\,|^\gamma\,sign(s_x) - \delta f(x,t) - d_x(t)$$

$$+ D^{2\alpha}x_d - D^{\alpha-1}\left[\left[k_{x1}(|\,e_{x1}\,| + |\,e_{x1}\,|^\beta) + k_{x2}.\right.\right.$$

$$\left.\left.(|\,e_{x2}\,| + |\,e_{x2}\,|^\beta) + (sign(e_{x1})D^{1-\alpha}e_{x2})\right]sign(e_{x2})\right]$$

(22)

For rejecting the model uncertainties and disturbances the control law is modified to,

$$u_{x1} = -k_{x3}s_x - k_{x4} \mid s_x \mid^\gamma sign(s_x) - (\delta_{x1} + \delta_{x2})sign(s_x)$$
$$+ D^{2\alpha}x_d - D^{\alpha-1}\left[\left[k_{x1}(\mid e_{x1} \mid + \mid e_{x1} \mid^\beta) + k_{x2}.\right.\right.$$
$$\left.\left.(\mid e_{x2} \mid + \mid e_{x2} \mid^\beta) + (sign(e_{x1})D^{1-\alpha}e_{x2})\right]sign(e_{x2})\right]$$

(23)

where δ_{x1} and δ_{x2} are positive tuning parameters. Like wise control laws for rest of the position and Euler angels can be calculated. Control law for y position tracking is:

$$u_{y1} = -k_{y3}s_y - k_{y4} \mid s_y \mid^\gamma sign(s_y) - (\delta_{y1} + \delta_{y2})sign(s_y)$$
$$+ D^{2\alpha}y_d - D^{\alpha-1}\left[\left[k_{y1}(\mid e_{y1} \mid + \mid e_{y1} \mid^\beta) + k_{x2}\right.\right.$$
$$\left.\left.k_{y2}(\mid e_{y2} \mid + \mid e_{y2} \mid^\beta) + (sign(e_{y1})D^{1-\alpha}e_{y2})\right]sign(e_{y2})\right]$$

Control law for altitude z tracking is:

$$u_{z1} = -k_{z3}s_z - k_{z4} \mid s_z \mid^\gamma sign(s_z) - (\delta_{z1} + \delta_{z2})sign(s_z)$$
$$+ D^{2\alpha}z_d - D^{\alpha-1}\left[\left[k_{z1}(\mid e_{z1} \mid + \mid e_{z1} \mid^\beta) + k_{z2}.\right.\right.$$
$$\left.\left.(\mid e_{z2} \mid + \mid e_{z2} \mid^\beta) + (sign(e_{z1})D^{1-\alpha}e_{z2})\right]sign(e_{z2})\right]$$

Control law for roll ϕ tracking is:

$$u_\phi = J_x\left(- k_{\phi3}s_\phi - k_{\phi4} \mid s_\phi \mid^\gamma sign(s_\phi) - \dot{\theta}\dot{\psi}(\frac{J_y - J_z}{J_x})\right.$$
$$- (\delta_{\phi1} + \delta_{\phi2})sign(s_\phi) + D^{2\alpha}\phi_d - D^{\alpha-1}\left[k_{\phi1}(\mid e_{\phi1}) \mid\right.$$
$$+ \mid e_{\phi1}) \mid^\beta) + k_{\phi2}(\mid e_{\phi2}) \mid + \mid e_{\phi2}) \mid^\beta) + (sign(e_{\phi1})\times$$
$$\left.\left.D^{1-\alpha}e_{\phi2})\right]sign(e_{\phi2})\right]\right)$$

Control law for pitch θ is:

$$u_\theta = J_y \bigg(-k_{\theta3}s_\theta - k_{\theta4} \mid s_\theta \mid^\gamma sign(s_\theta) - \dot\phi\dot\psi(\frac{J_z - J_x}{J_y})$$

$$-(\delta_{\theta1} + \delta_{\theta2})sign(s_\theta) + D^{2\alpha}\theta_d - D^{\alpha-1}\bigg[\bigg[k_{\theta1}(\mid e_{\theta1} \mid$$

$$+\mid e_{\theta1} \mid^\beta) + k_{\theta2}(\mid e_{\theta2} \mid + \mid e_{\theta2} \mid^\beta) + (sign(e_{\theta1})\times$$

$$D^{1-\alpha}e_{\theta2})\bigg]sign(e_{\theta2})\bigg]\bigg)$$

Control law for yaw θ is:

$$u_\psi = J_z \bigg(-k_{\psi3}s_\phi - k_{\psi4} \mid s_\psi \mid^\gamma sign(s_\psi) - \dot\phi\dot\theta(\frac{J_x - J_y}{J_z})$$

$$-(\delta_{\psi1} + \delta_{\psi2})sign(s_\psi) + D^{2\alpha}\psi_d - D^{\alpha-1}\bigg[\bigg[k_{\psi1}(\mid e_{\psi1} \mid$$

$$+\mid e_{\psi1} \mid^\beta) + k_{\psi2}(\mid e_{\psi2}) \mid + \mid e_{\psi2}) \mid^\beta) + (sign(e_{\psi1})\times$$

$$D^{1-\alpha}e_{\psi2}\bigg]sign(e_{\psi2})\bigg]\bigg)$$

6 Stability Analysis of Controller

For the stability analysis, two different Lyapunov function have been taken where one is for reaching phase stability analysis and the other is for sliding phase.

6.1 Reaching Phase Stability

For reaching phase we have to show that reachability law $\dot s_x(t) = -k_{x3}s_x - k_{x4} \mid s_x \mid^\gamma sign(s_x)$ converges to zero. Let us take Lyapunov candidate as:

$$V_r = \mid s_x \mid \tag{24}$$

The derivative of V_r is

$$\dot V_r = sign(s_x)\dot s_x \tag{25}$$

Substituting derivative of sliding surface using Eq. (17) in Eq. (25)

$$\dot V_r = \bigg(D^\alpha e_{x2} + D^{\alpha-1}\bigg[\bigg[k_{x1}(\mid e_{x1}) \mid + \mid e_{x1}) \mid^\beta) + k_{x2}($$

$$\mid e_{x2} \mid + \mid e_{x2}) \mid^\beta) + (sign(e_{x1})D^{1-\alpha}e_{x2})\bigg]sign(e_{x2})\bigg]\bigg)$$

$$\times sign(s_x)$$

$$\tag{26}$$

After substituting the value of $D^\alpha e_{x2}$ from error dynamics and inserting the control law u_{x1} Eq. (26) will become:

$$\dot{V}_r \leq -sign(s_x)(k_{x3}s_x + k_{x4} \mid s_x \mid^\gamma sign(s_x)) \tag{27}$$

Using $sign(s_x)s_x = \mid s_x \mid$ and $sign^2(s_x)s_x = 1$. The Lyapunov derivative will become-

$$\dot{V}_r \leq -k_x(\mid s_x \mid + \mid s_x \mid^\gamma) \leq -k_x \mid s_x \mid \tag{28}$$

where, $k_x = min(k_{x3}, k_{x4})$. Equation (28) is negative definite hence it is stable.

6.2 Sliding Phase Stability

for sliding phase stability we have to show that both the error states of x position tracking converges to zero. Let us take Lyapunov candidate as:

$$V_s = \mid e_{x1} \mid + \mid e_{x2} \mid \tag{29}$$

The derivative of V_s is

$$\dot{V}_s = sign(e_{x1})\dot{e}_{x1} + sign(e_{x2})\dot{e}_{x2} \tag{30}$$

Equation (25) can also be written as using property of fractional order theory

$$\dot{V}_s = sign(e_{x1})[D^{1-\alpha}D^\alpha e_{x1}] + sign(e_{x2})[D^{1-\alpha}D^\alpha e_{x2}]$$

From Eq. (10) and Eq. (18) substituting the values of $D^\alpha e_{x1}$ and $D^\alpha e_{x2}$ in the above Eq.

$$
\begin{aligned}
\dot{V}_s = {} & sign(e_{x1})[D^{1-\alpha}e_{x2}] - sign(e_{x2})D^{1-\alpha}D^{\alpha-1} \times \\
& \left[\Big[k_{x1}(\mid e_{x1}) \mid + \mid e_{x1}) \mid^\beta) + k_{x2}(\mid e_{x2}) \mid + \mid e_{x2}) \mid^\beta) \right. \\
& \left. + (sign(e_{x1})D^{1-\alpha}e_{x2}) \Big] sign(e_{x2}) \right]
\end{aligned}
\tag{31}
$$

After simplifying Eq. (31), we get

$$\dot{V}_s = -\Big[k_{x1}(\mid e_{x1}) \mid + \mid e_{x1} \mid^\beta) + k_{x2}(\mid e_{x2}) \mid + \mid e_{x2}) \mid^\beta \Big]$$

which is negative definite. Hence both the reaching phase and sliding phase of designed fractional order controller is stable.

7 Adaptive Neural Network Based Controller Design

7.1 Function Approximation

The activation function used for the approximation is of the following form:

$$S(P) = exp\left[-\frac{(P - \mu_j)^T(P - \mu_j)}{\sigma^2} \right], j = 1, 2, ..., m \tag{32}$$

where, μ_j is the position of the hidden node, P is the input vector, σ is the width of the Gaussian function. The function is approximated by the product of optimized weights with activation function and is written as:

$$F(P) = W^T S(P) \tag{33}$$

where, W is the weight vector to be tuned optimally. The estimated error of weight $\tilde{W}$ is defined as:

$$\tilde{W} = W - \hat{W} \tag{34}$$

7.2 Tuning of Weights Using Optimality Conditions

Following objective function is considered for minimizing the weights

$$J(\mu) = \sum_{j=1}^{m} \sum_{i=1}^{n} \|P_i^{(j)} - \mu_j\|^2 \tag{35}$$

where, $\|P_i^{(j)} - \mu_j\|^2$ is a squared Euclidean distance between cluster center μ_j a data point $P_i^{(j)}$ and the corresponding cluster center μ_j. The optimal values are obtained in the following steps:

- Randomly select the initial m centres which are close to zero, $\mu = \{\mu_1, \mu_2, ..., \mu_m\}$.
- Then choose the minimum center μ_j which is close to the data point Z_i.
- Continuously update the center for each cluster and repeat the process untill μ converges.

It should be noted that the optimal distribution of hidden nodes improves the approximation ability of neural network compared to lattice distribution of nodes.

7.3 Adaptive Neural Network Based Control Law

Let us again consider the Eq. (10) and modify the control law using neural network based estimator,

$$u_{x1} = -k_{x4}D^\alpha e_{x1} + D^{2\alpha}x_d - \hat{W}S(X) \tag{36}$$

where, $\hat{W}S(X)$ is the estimated value of function $\delta f(x, t) + d_x(t)$. Likewise, the control law for y position and z position will be,

$$u_{y1} = -k_{y4}D^\alpha e_{y1} + D^{2\alpha}y_d - \hat{W}S(Y) \tag{37}$$

$$u_{z1} = -k_{z4}D^\alpha e_{z1} + D^{2\alpha}z_d - \hat{W}S(Z) \tag{38}$$

where, $\hat{W}S(Y)$ is the estimated value of function $\delta f(y, t) + d_y(t)$ and $\hat{W}S(Z)$ is the estimated value of function $\delta f(z, t) + d_z(t)$. Now, for roll, pitch and yaw the proposed neural network based controllers are

$$u_{\phi1} = -k_{\phi4}D^\alpha e_{\phi1} + D^{2\alpha}\phi_d - \hat{W}S(\phi) \tag{39}$$

$$u_{\theta 1} = -k_{\theta 4} D^\alpha e_{\theta 1} + D^{2\alpha} \theta_d - \hat{W} S(\theta) \tag{40}$$

$$u_{\psi 1} = -k_{\psi 4} D^\alpha e_{\psi 1} + D^{2\alpha} \psi_d - \hat{W} S(\psi) \tag{41}$$

where, $\hat{W} S(\phi)$ is the estimated value of function $\dot{\theta}\dot{\psi}(\frac{J_y - J_z}{J_x}) + \delta f(\phi, t) + d_\phi(t)$ and $\hat{W} S(\theta)$ is the estimated value of function $\dot{\phi}\dot{\psi}(\frac{J_z - J_x}{J_y}) + \delta f(\theta, t) + d_\theta(t)$ and $\hat{W} S(\psi)$ is the estimated value of function $\dot{\phi}\dot{\theta}(\frac{J_x - J_y}{J_z}) + \delta f(\psi, t) + d_\psi(t)$.

Stability Analysis. In this section only the stability of x position controller is given. Let us take the Lyapunov function as;

$$V = |e_{x1}| + |e_{x2}| + \frac{1}{\beta} \tilde{W}^2 \tag{42}$$

Taking the derivative of Eq. (42) The derivative of V_s is

$$\dot{V}_s = sign(e_{x1})\dot{e}_{x1} + sign(e_{x2})\dot{e}_{x2} - \frac{2}{\beta} \tilde{W} \dot{\hat{W}} \tag{43}$$

Equation (43) can also be written as using property of fractional order theory

$$\dot{V}_s = sign(e_{x1})[D^{1-\alpha} D^\alpha e_{x1}] + sign(e_{x2})[D^{1-\alpha} D^\alpha e_{x2}] - \frac{2}{\beta} \tilde{W} \dot{\hat{W}} \tag{44}$$

Now, substituting the value of $D^\alpha e_{x1}$ and $D^\alpha e_{x2}$ in Eq. (44)

$$\dot{V}_s = sign(e_{x1})[D^{1-\alpha} e_{x2}] + sign(e_{x2})[D^{1-\alpha}(\delta f(x,t) + d_x(t) + u_{x1} - D^{2\alpha} x_d)] - \frac{2}{\beta} \tilde{W} \dot{\hat{W}} \tag{45}$$

Now, substituting the value of u_{x1} in Eq. (45)

$$\dot{V}_s = sign(e_{x1})[D^{1-\alpha} e_{x2}] + sign(e_{x2})[D^{1-\alpha}(\delta f(x,t) + d_x(t) - k_{x4} D^\alpha e_{x1} + D^{2\alpha} x_d \\ - \hat{W} S(X) - D^{2\alpha} x_d)] - \frac{2}{\beta} \tilde{W} \dot{\hat{W}} \tag{46}$$

Since, $\delta f(x,t) + d_x(t) = W S(X)$, substituting the value in Eq. (47)

$$\dot{V}_s = sign(e_{x1})[D^{1-\alpha} e_{x2}] + sign(e_{x2})[D^{1-\alpha} W S(X) - k_{x4} D^\alpha e_{x1} + D^{2\alpha} x_d \\ - \hat{W} S(X) - D^{2\alpha} x_d] - \frac{2}{\beta} \tilde{W} \dot{\hat{W}} \tag{47}$$

Now, further simplifying

$$\dot{V}_s \leq |e_{x1}| - |e_{x2}| + \tilde{W}[D^{1-\alpha} S(X) - \frac{2}{\beta} \dot{\hat{W}}] \tag{48}$$

If the adaptive tuning law will be selected such that $\dot{\hat{W}} = \frac{\beta D^{1-\alpha} S(x)}{2}$, then the system will be asymptotically stable.

Likewise adaptive tuning laws for other positions and Euler angles can be calculated.

8   Simulation Results

Proposed approach has been validated in MATLAB where following set of parameters specification has been taken as:

Mass (m)	0.2 Kg
Inertia (I_{xx})	1.676×10^{-2}
Inertia (I_{yy})	1.676×10^{-2}
Inertia (I_{zz})	2.314×10^{-2}

The tuning parameters are selected as. $k_{x1} = k_{x2} = 1.5, k_{y1} = k_{y2} = 1.75, k_{z1} = k_{z2} = 1.75$ and $k_{\phi1} = k_{\phi} = 1.0, k_{\theta} = k_{\theta} = 1.0, k_{\psi} = k_{\psi} = 1.2$. The fractional derivatives are selected as $\alpha = 0.65$ and $\beta = 0.0.45$. The proposed approach has been validated for two cases i.e. quadrotor hovering at 1.25 m and circular and spiral shaped trajectory tracking. For hovering case fractional order controller is used and for trajectory tracking adaptive neural network based fractional controller is used.

8.1   Quadrotor Hovering

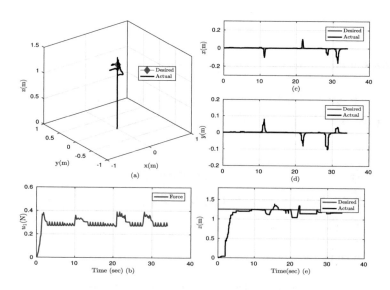

Fig. 4. Hovering of UAV using only fractional order sliding mode control.

Here, the objective is to takeoff the quadrotor to $z_d = 1.25$ m and hover thereafter at this height. To check the robustness of the proposed controller, a disturbance has been added at hovering in all of the three directions x, y and z at different time instants, the maximum bound on the magnitude of disturbances in all the three directions are $0.2 * sint$. Disturbances are applied in all the three directions at different instants of time to check the robustness of the controller. The simulation results are shown in Fig. 4 where we see that the quadrotor successfully reaches at 1.25 m and hover as shown in Fig. 4(a) in spite of the disturbance. Thus we can conclude from Fig. 4 that the quadrotor effectively counteract the disturbance and hover continuously at 1.25 m. From the result of Fig. 4(b) one can observe that control effort are not very smooth. The controller used in the hovering case has been proposed in [12].

8.2 Circular and Spiral Shaped Trajectory Tracking

Now, the quadrotor is required to track the circular and spiral shaped trajectory which is generated as follows by desired positions as: For circular shape-$x_d = 01 * sin(0.15 * t)$, $y_d = 2 * cos(0.2 * t)$ and $z_d = 1.5$, For spiral shaped-$x_d = 01 * sin(0.15 * t)$, $y_d = 2 * cos(0.2 * t)$ and $z_d = 1.5 * t$. In this case adaptive neural network based fractional order controller is applied. The proposed approach is the modification of the [12] and obtained results are shown in Fig. 5 and Fig. 6. From Fig. 5(a) and Fig. 6(a), we see that trajectory tracking performance is smooth as well as control effort is also reduced significantly compared to only fractional order control effort. Hence for real time applications it is more convenient to apply proposed neural network based controller compared to existing one [12].

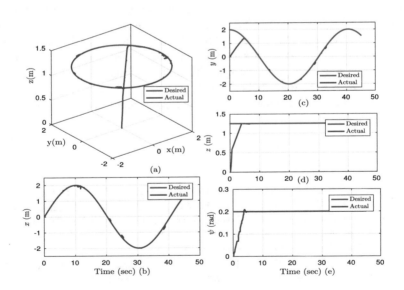

Fig. 5. UAV tracking circular shaped trajectory using adaptive fractional order controller.

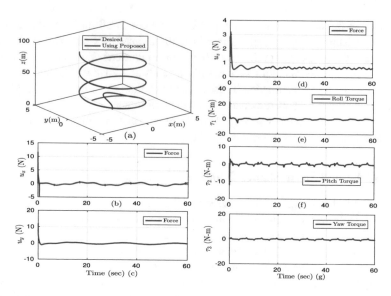

Fig. 6. UAV tracking spiral shaped trajectory using proposed fractional order controller.

9   Conclusion and Future Scope

In the proposed work a novel adaptive neural network based fractional order controller is designed for fractional order model of the quadrotor. Both the fractional order model of the quadrotor and fractional order adaptive neural network controller achieves better results compared to integer order system and controller. The nodes of the neural network are selected optimally. The comparative study has been done with the proposed method with the existing fractional order sliding mode controller [12]. The future scope of the present work is to design a fractional order multiagent model of a quadrotor with adaptive learning rate.

References

1. Aubry, M., Maturana, D., Efros, A.A., Russell, B.C., Sivic, J.: Seeing 3D chairs: exemplar part-based 2D-3D alignment using a large dataset of cad models. In: Proceedings of the IEEE Conference on Computer Vision and Pattern Recognition, pp. 3762–3769 (2014)
2. Boiko, I., Fridman, L.: Analysis of chattering in continuous sliding-mode controllers. IEEE Trans. Autom. Control **50**(9), 1442–1446 (2005)
3. Cajo, R., Mac, T.T., Plaza, D., Copot, C., De Keyser, R., Ionescu, C.: A survey on fractional order control techniques for unmanned aerial and ground vehicles. IEEE Access **7**, 66864–66878 (2019)
4. Chen, L., Saikumar, N., HosseinNia, S.H.: Development of robust fractional-order reset control. IEEE Trans. Control Syst. Technol. **28**(4), 1404–1417 (2019)
5. Cohen, M.R., Abdulrahim, K., Forbes, J.R.: Finite-horizon LQR control of quadrotors on $SE\_2(3)$. IEEE Robot. Autom. Lett. **5**(4), 5748–5755 (2020)
6. Giusti, A., et al.: A machine learning approach to visual perception of forest trails for mobile robots. IEEE Robot. Autom. Lett. **1**(2), 661–667 (2015)

7. He, W., Dong, Y.: Adaptive fuzzy neural network control for a constrained robot using impedance learning. IEEE Trans. Neural Netw. Learn. Syst. **29**(4), 1174–1186 (2017)
8. Hua, C., Chen, J., Guan, X.: Fractional-order sliding mode control of uncertain QUAVs with time-varying state constraints. Nonlinear Dyn. **95**(2), 1347–1360 (2019). https://doi.org/10.1007/s11071-018-4632-0
9. Izaguirre-Espinosa, C., Muñoz-Vázquez, A.J., Sánchez-Orta, A., Parra-Vega, V., Fantoni, I.: Fractional-order control for robust position/yaw tracking of quadrotors with experiments. IEEE Trans. Control Syst. Technol. **27**(4), 1645–1650 (2018)
10. Labbadi, M., Nassiri, S., Bousselamti, L., Bahij, M., Cherkaoui, M.: Fractional-order fast terminal sliding mode control of uncertain quadrotor UAV with time-varying disturbances. In: 2019 8th International Conference on Systems and Control (ICSC), pp. 417–422. IEEE (2019)
11. Liu, H., Li, D., Zuo, Z., Zhong, Y.: Robust three-loop trajectory tracking control for quadrotors with multiple uncertainties. IEEE Trans. Ind. Electron. **63**(4), 2263–2274 (2016)
12. Maurya, H., Singh, P., Yogi, S., Behera, L., Verma, N.: Fractional order tracking control of unmanned aerial vehicle in presence of model uncertainties and disturbances. In: Proceedings of the 18th International Conference on Informatics in Control, Automation and Robotics-ICINCO, INSTICC, pp. 274–281. SciTePress (2021)
13. Mogili, U.R., Deepak, B.: Review on application of drone systems in precision agriculture. Procedia Comput. Sci. **133**, 502–509 (2018)
14. Odibat, Z.: Approximations of fractional integrals and Caputo fractional derivatives. Appl. Math. Comput. **178**(2), 527–533 (2006)
15. Oliva-Palomo, F., Muñoz-Vázquez, A.J., Sánchez-Orta, A., Parra-Vega, V., Izaguirre-Espinosa, C., Castillo, P.: A fractional nonlinear PI-structure control for robust attitude tracking of quadrotors. IEEE Trans. Aerosp. Electron. Syst. **55**(6), 2911–2920 (2019)
16. Ríos, H., Falcón, R., González, O.A., Dzul, A.: Continuous sliding-mode control strategies for quadrotor robust tracking: real-time application. IEEE Trans. Ind. Electron. **66**(2), 1264–1272 (2018)
17. Shi, X., et al.: Adaptive fractional-order SMC controller design for unmanned quadrotor helicopter under actuator fault and disturbances. IEEE Access **8**, 103792–103802 (2020)
18. Singh, P., Gupta, S., Behera, L., Verma, N.K., Nahavandi, S.: Perching of nano-quadrotor using self-trigger finite-time second-order continuous control. IEEE Syst. J. **15**(4), 4989–4999 (2020)
19. Singh, P., Nandanwar, A., Behera, L., Verma, N.K., Nahavandi, S.: Uncertainty compensator and fault estimator-based exponential supertwisting sliding-mode controller for a mobile robot. IEEE Trans. Cybern. **52**(11), 11963–11976 (2021)
20. Sun, C., Gao, H., He, W., Yu, Y.: Fuzzy neural network control of a flexible robotic manipulator using assumed mode method. IEEE Trans. Neural Netw. Learn. Syst. **29**(11), 5214–5227 (2018)
21. Wang, H., Ye, X., Tian, Y., Zheng, G., Christov, N.: Model-free-based terminal SMC of quadrotor attitude and position. IEEE Trans. Aerosp. Electron. Syst. **52**(5), 2519–2528 (2016)
22. Zhou, W., Zhu, P., Wang, C., Chu, M.: Position and attitude tracking control for a quadrotor UAV based on terminal sliding mode control. In: 2015 34th Chinese Control Conference (CCC), pp. 3398–3404. IEEE (2015)
23. Xiong, J.J., Zhang, G.: Sliding mode control for a quadrotor UAV with parameter uncertainties. In: 2016 2nd International Conference on Control, Automation and Robotics (ICCAR), pp. 207–212. IEEE (2016)
24. Yu, G., Cabecinhas, D., Cunha, R., Silvestre, C.: Nonlinear backstepping control of a quadrotor-slung load system. IEEE/ASME Trans. Mechatron. **24**(5), 2304–2315 (2019)

Design and Development of a Dexterous Teleoperation Setup for Nuclear Waste Remote Manipulation

Florian Gosselin[✉] [iD] and Mathieu Grossard

Université Paris-Saclay, CEA, LIST, 91120 Palaiseau, France
florian.gosselin@cea.fr

Abstract. Teleoperation is entering a new era. Despite still in use today in the nuclear industry, simple purely mechanical or robotic 6 degrees of freedom (DoF) master-slave systems equipped with bi-digital grippers on the slave side and simple handles on the master side are not able to answer the challenge of remote manipulation at scale. These historical remotely controlled robotic solutions, inherited from mechanical master slave systems developed decades ago for research activities performed in gloveboxes and hot cells, allow operators an efficient and safe access to dangerous materials at distance. They are adapted when the variety of the to-be-manipulated objects remains limited, especially when these objects can be adapted for remote manipulation. They are however no more sufficient when one has to handle a much higher quantity of much more diverse objects, as it is typically the case when processing nuclear waste accumulated in huge quantities over time and/or produced at the occasion of nuclear power plants' dismantling operations. The quantity and diversity of nuclear waste require more efficient and versatile systems. To answer this challenge and increase the operators' productivity, we developed a complete bimanual teleoperation setup able to perform remote dexterous manipulation tasks. This article describes the hardware and software architecture of this platform, which is notably composed of a novel dexterous master-slave system combining a tri-digital master glove and a remotely controlled three fingers gripper. Both make use of highly backdrivable actuators and transmissions, and the proposed coupling schemes allow intuitive control of various grasp types. As a result, the proposed setup enables dexterous and force-sensitive control, and it is well-suited for high fidelity force-reflection tele-presence.

Keywords: Teleoperation · Dexterous manipulation · Multi-finger gripper · Hand exoskeleton · Force feedback

1 Introduction

The rise of the nuclear industry in the middle of the twentieth century required the development of efficient processes allowing to handle radioactive material without exposing operators to danger. The solution found by researchers and engineers to grasp and manipulate radioactive objects in laboratory settings was to use remote manipulation means, among which telemanipulators are the most advanced and efficient solutions, to access safely dangerous materials disposed in gloveboxes and hot cells. The objects themselves, in limited number, were adapted for remote manipulation, allowing their grasping and manipulation with simple 6 DoF master-slave systems equipped with bi-digital grippers. Various systems of this kind, being either purely mechanical of robotic devices, were developed and used over years in nuclear installations [1, 2]. They prove to be very efficient, especially those benefiting from computer assisted telemanipulation functions, and are still in use today, for example in the recycling plant of La Hague in France [3, 4].

Such solutions are however not able to answer the challenge of remote manipulation at scale. They were perfectly adapted for handling few radioactive objects, but they are no more sufficient when one has to handle a high quantity of very various objects, as it is typically the case when processing nuclear waste accumulated in huge quantities over the years of exploitation of nuclear power plants and/or produced at the occasion of their dismantling. The quantity and diversity of nuclear waste materials require more efficient and versatile systems. The objective of the RoMaNS (i.e. Robotic Manipulation for Nuclear Sort and Segregation) project, financed by the European Horizon 2020 research program, was to develop novel solutions able to answer this challenge [5, 6]. To better understand the problem, it can be recalled that only highly irradiated material must be stored in high-level storage containers and facilities which are extremely expensive and resource intensive. On the contrary, low level waste can be placed in low-level storage containers. In practice however, waste of mixed contamination levels were sometimes put together in storage containers, especially in the older nuclear sites (some dates back to the 1940s). It is now time to clean up this waste stock and develop a more sustainable solution to store each waste item in an adequate container. The vast quantities of legacy nuclear waste (in the UK for example, intermediate level waste amount to about 1.4 million cubic meters [7]) makes it critical to advance the state of the art in telemanipulation in order to solve the safety-critical industrial problem of sorting and segregating irradiated material.

This sorting process requires opening thousands of legacy waste containers, extracting their potentially very various contents (pieces of fuel rod casing, contaminated tools and rubble, irradiated suits, rubber gloves, etc.), and sorting and segregating the most highly contaminated objects. The high level of radiations of some waste material prevents manual operation, and the process can only be performed using remotely controlled robots. As previously explained, state-of-the-art 6 DoF master-slave systems equipped with bi-digital grippers are not a viable solution therefore in the longterm due to some important limitations. First, they do not allow grasping all kinds of objects being present in the containers. Also, they must be under permanent direct control of human operators, with poor productivity. The aim of the ROMANS project was to develop more dexterous and versatile telemanipulation means and mixed autonomy solutions allowing to answer these limits.

This paper, which is an extended version of the work presented in [8], addresses the first issue. Compared to [8], which focused on the master hand device, this article introduces additional information on the whole bi-manual teleoperation platform fitted for waste processing, with details on the slave hand and dexterous master-slave coupling schemes. It is organized as follows: Sect. 2 first shortly introduces the whole platform, then Sect. 3 and 4 present the dexterous slave gripper and the master hand device. Sections 5 and 6 give details on the coupling schemes used to control the system and introduce the first validation tests, and Sect. 7 concludes the paper.

2   Bi-manual Teleoperation Platform

Figure 1 illustrates the bi-manual teleoperation set-up. This platform is composed of two slave robots in order to allow for a natural operation with both hands, a Stäubli RX160 equipped with a dexterous gripper and a prototype cobot developed at CEA equipped with a classical bi-digital gripper. These robots are controlled using two input devices mapping the slave-arm capabilities. A first Virtuose 6D TAO (www.haption. com) equipped with our novel hand master glove is used for the control of the first robot, allowing fine control of the dexterous gripper with force feedback on both the palm and fingers, and a second Virtuose 6D TAO equipped with a handle is used to control the second robot.

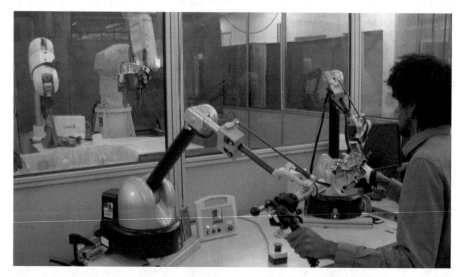

Fig. 1. Bi-manual dexterous teleoperation set-up (source [8]).

It is worth noting that to efficiently perform dexterous operations, the human beings often make use of both hands. This configuration allows some level of parallelization, and most importantly to concurrently perform complementary operations (e.g. holding a container with one hand and opening the lid with the second hand, opening a container and grasping an object inside it, grasping an object and making an operation on it, etc.). This configuration was used here, with one robot carrying a simple gripper used for rough operations, and the second one equipped with the three fingers gripper allowing to perform more dexterous tasks.

All components of this platform share a high backdrivability obtained either by design (esp. regarding actuators and transmissions) for the master arms, dexterous master glove and slave hand, and cobot prototype, or using a force/torque sensor for the Stäubli robot. As a result, both robots enable teleoperation with high fidelity force-reflection, thus intuitive and efficient remote operation. Further details on the slave and master hands are given below.

3 Dexterous Three Fingers Gripper

3.1 Design Rationales and Principles

Human operators prove to be very efficient for waste sorting, comfort and safety issues aside. Their large experience of handling various objects in the daily life associated with the formidable flexibility, dexterity and sensitivity of the human hand, allows them to quickly adapt to any situation. The human hand has thus been an important source of inspiration for the design of dexterous multi-fingered robotic grippers like for example the Shadow Robot hand [9, 10], the anthropomorphic CEA-LIST hand [11] or the DLR AWIWI hand [12]. Some of these devices are natively backdrivable, allowing to estimate the grasping forces without additional force or torque sensors, while others make use of tendon tension sensors, force/torque sensors or tactile sensors therefore. Such grippers are capable of reproducing grasps found in usual grasping taxonomies [13, 14] and they allow performing dexterous tasks under direct human control in telerobotics with force feedback [15]. Their design, as well as their control, are however very complex and they remain in practice limited to laboratory experiments. None of them reached industrial settings and teleoperation slave arms and industrial robots still most often make use of simple bi-digital grippers or dedicated tools which in turn suffer a poor versatility.

A compromise between these two extremes is necessary to overcome this situation. As the most bulky and heavy components in a robot gripper are its actuators, a logical solution is to reduce their number. This can be done in two ways that we combine here:

- First, the number of fingers can be reduced. As we focus here on grasping, a configuration with three fingers is sufficient (should we have also targeted manipulation, a fourth finger as on the ROBIOSS hand [16] or additional embedded actuation means as in [17] would have been necessary). Yet to better adapt the gripper's configuration to the shape of the grasped objects and improve stability, a reconfigurable palm as present for example in the BarretHand [18] is implemented.
- Second, the gripper can be under-actuated, with coupling rods allowing to synchronize several degrees of freedom and springs allowing to passively adapt to the grasped objects' geometry as proposed in [19] and [20].

3.2 Electro-mechanical Design

The three-fingers robotic gripper is illustrated in Fig. 2. Following the above-mentioned principles, this hand is composed of three fingers, one of which is fixed and the other two being mobile in abduction-adduction so as to allow cylindrical grasps when the adduction is null, spherical grasps around the center of the abduction-adduction range of motion and planar grasps when fingers are fully adducted (see Fig. 5 in Sect. 3.4 below). Delrin gears actuated by a single motor (hidden in Fig. 2) allow coupling the abduction-adduction movements of the mobile fingers and the palm facing the fingers is slightly curved to favor the stability of the grasps once the objects are held in the gripper.

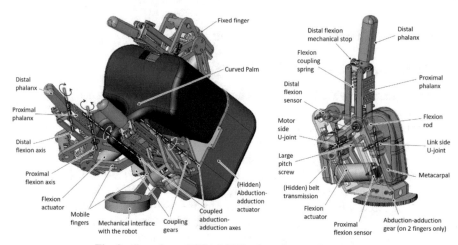

Fig. 2. Overview of CEA LIST's three fingers slave hand.

All three fingers have similar kinematics above the abduction-adduction axis. They are composed of a first link equivalent to human metacarpals, a proximal and a distal phalange. The fingers are underactuated, i.e. there is only one actuator controlling the two flexion DoFs. The actuation system is composed of a Maxon DCX26L DC motor (48 V, continuous torque 59.1 mNm, peak torque 697 mNm [21]) associated with a belt transmission (reduction ratio equal to 1) driving a highly backdriveable ball screw (neutral diameter 5 mm, pitch 2 mm) which itself drives a nut undergoing a translation movement. At the output of the system, the nut drives a rod disposed at the base of the proximal phalange. The principle is the same as in Screw Cable Systems [22], except a direct transmission without any cable is used here. This solution has the advantage of a higher efficiency, at the price of a non constant reduction ratio as the lever arm between the driving screw-nut system and the moving rod axis varies as the finger moves in flexion. The finger itself is composed of two phalanges whose movements are coupled with a four bar mechanism and a spring. As long as there is no contact on the proximal phalange, the finger remains straight. The distal phalanx bends only when a contact occurs on the proximal phalanx, with a triggering force regulated by the four bar geometry and the spring characteristics (this force is equal to 30N here). One benefit of underactuation is that it allows both precision grasps when the object is contacted with the distal phalanges and power grasps of various objects when the contact first occurs on the proximal phalanges. In the latter case, the grasping movement has two phases: a sweeping phase, where the proximal links contact the object, and a caging phase, where the distal links make contact to fully enclose the object. It is worth noting that if contact on any finger is obstructed, the others continue to move, until the hand fully envelopes the object in a power grasp. Underactuation (both within the individual fingers and between the fingers) is thus favorable for achieving form closure as it can aid in maximizing the grasp contact surfaces. Moreover, all fingers are provided with interchangeable pads that can be attached on the proximal and distal phalanges so as to adapt the grasps form factor and the friction coefficient. Owing to the specifications of the different components, the continuous force in the nut can be estimated at about 167 N with an hypothesis of a 90% efficiency (i.e. peak force 1970 N). This corresponds to a theoretical flexion torque equal to 4.9 Nm continuous and 58 Nm peak and a maximum force of about 33 N continuous and 387 N peak at the fingertip when the finger is fully extended and normal to the palm as shown in Fig. 2 (lever arm equal to 29.5 mm, distance between the flexion rod axis and the fingertip equal to about 150 mm).

The principle is the same on the abduction-adduction axis, except that an indirect actuation similar to a SCS is used, with a belt replacing the cable however. The actuators and screw and nut systems are the same as on the flexion axes, but the primary belt differs in that it introduces a 2.33 reduction ratio, hence a force of 390 N continuous and about 4600 N peak in the nut. This force is equally distributed on the two abduction-adduction axes, whose movements are coupled using Delrin gears, through output pulleys whose neutral diameter equals 29 mm, hence a theoretical continuous torque of 5.7 Nm (peak torque 67 Nm).

All four actuators are equipped with 1024 ppt Maxon ENX16 magneto-optical encoders. Additional Megatron potentiometers are used to sense the rotation around the distal flexion axes.

3.3 Kinematics

The simplified kinematic model of the three fingers gripper is illustrated in Fig. 3. It makes use of Denavit Hartenberg notations [23]. For each finger i, rotation θ_{i1} allows abduction-adduction, θ_{i2} proximal flexion and θ_{i3} distal flexion. A frame $R_{ik} = (O_{ik}, X_{ik}, Y_{ik}, Z_{ik})$ is first associated with each link k of finger i. By denoting d_{ik} (resp. α_{ik}) the distance from (resp. the angle between) $Z_{i\,k-1}$ to (and) Z_{ik} along axis $X_{i\,k-1}$ and r_{ik} and θ_{ik} the distance from (resp. the angle between) $X_{i\,k-1}$ to (and) X_{ik} along axis Z_{ik}, we can write the transformation matrix from $R_{i\,k-1}$ to R_{ik} as follows:

$$T_{i\,k-1\,k} = \text{rot}(X_{i\,k-1}, \alpha_{ik}).\text{trans}(X_{i\,k-1}, d_{ik}).\text{rot}(Z_{ik}, \theta_{ik}).\text{trans}(Z_{ik}, r_{ik}) \quad (1)$$

Additional frames $R_{iT} = (O_{iT}, X_{iT}, Y_{iT}, Z_{iT})$ are introduced at the tip of each finger i, so as to allow taking into account different lengths and shapes of the fingertips. By denoting d_{iT} the length of the distal phalanx, we have:

$$T_{i\,3T} = \text{rot}(X_{i3}, \alpha_{iT}).\text{trans}(X_{i3}, d_{iT}) \quad (2)$$

Finally, an additional frame $R_G = (O_G, X_G, Y_G, Z_G)$ is introduced at the base of the gripper (i.e. at level of the interface with the robot), allowing to express the configuration of the basis of all fingers in a common frame as follows:

$$T_{i\,G0} = \text{rot}(X_G, \alpha_{iG}).\text{trans}(Y_G, d_{iG}).\text{rot}(Z_{i0}, \theta_{iG}) \quad (3)$$

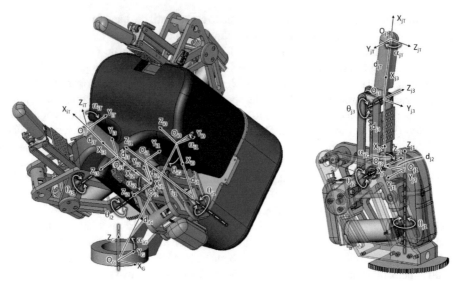

Fig. 3. Kinematics of the three fingers slave hand (left: complete gripper with index i used to describe the movable finger on the right, right: focus on the left finger labelled j).

The transformation matrix of each finger can then be written:

$$T_{i\ GT} = T_{i\ G0}.T_{i\ 01}.T_{i\ 12}.T_{i\ 23}.T_{i\ 3T} \tag{4}$$

In practice θ_{i1} is obtained from the abduction-adduction sensors and θ_{i2} and θ_{i3} from the flexion sensors and the actuation and four bar mechanisms closed loop equations.

3.4 Manufacturing and Integration

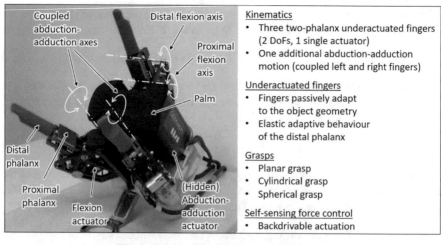

Fig. 4. CEA LIST's three fingers slave hand.

Figure 4 illustrates the prototype reconfigurable and under-actuated three-fingers robotic hand. The hand parts are made of aluminum and carbon fiber-reinforced plastics to remain as light as possible. All transmissions are placed inside the structural parts or covered with plastic shells in order to protect the mechanisms. Finally, the distal phalanges are covered with a soft polymeric envelope in order to increase grasping robustness when an object is held in hand, and a sharp end is provided in order to allow for precision grasps. The controller comes in a separate box integrating 24 V (for the sensors) and 48 V (for the actuators) power supplies, Beckhoff analog input modules and Ethercat couplers and Maxon EPOS3 Ethercat drives.

As can be seen in Fig. 5, the range of motion of the different axes allow for the generation of planar, spherical and cylindrical grasps. In extreme configurations, the movable fingers point towards each other, or towards the third and fixed finger.

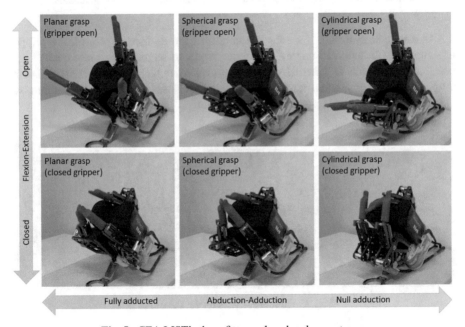

Fig. 5. CEA LIST's three fingers slave hand grasp types.

Such a three fingers design proves to be sufficient for coarse gripping and manipulating the to-be-sorted objects, yet it remains simple and rugged when compared to five-fingers hands which would be too fragile for such harsh environment. As shown in Fig. 6, it can generate power and precision grasp patterns [13], and it is capable of grasping a large variety of objects similar in size and weight to those encountered in nuclear waste containers.

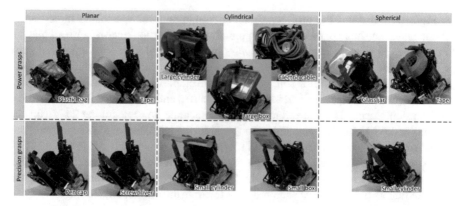

Fig. 6. Example grasps of representative objects as found in waste containers.

As can be seen in Fig. 7, during the course of the ROMANS project, the gripper was successively mounted on several robots thanks to its universal mechanical interface, among which a Yaskawa SIA10 (https://www.motoman.com/), an ABB IRB4600 (https://new.abb.com/) and a Stäubli RX160 (https://www.staubli.com/), for the purpose of integration, functional tests and controller tuning.

Fig. 7. Example set-ups on the slave side.

4 Tri-Digital Dexterous Hand Master

As mentioned in Sect. 1, most teleoperated robots used to date in nuclear sites remain master-slave systems equipped with simple handles and grippers developed decades ago. Fortunately, huge progress has been obtained in the meantime in dexterous force feedback robotics and VR haptics. The requirement for anthropomorphic devices able to assist humans in force demanding applications (e.g. military, civil security, firemen, and even industry) or to restore lost motor abilities (e.g. rehabilitation, disabled people assistance), as well as the rise of Virtual Reality applications, led to the development of numerous arm and hand orthoses, exoskeletons and master devices [24–28]. The lessons learnt from these works were taken into account for the development of the tri-digital hand master presented in this section.

4.1 Design Drivers

1/Dexterous Manipulation: the use of a tri-digital hand master appears as the most logical solution to control the three fingers slave hand. Interestingly, a more general study of manual interactions shows that three-fingers haptic devices offer a good compromise between manipulation capabilities and complexity [29]. As illustrated in Fig. 8, the percentage of our daily life activities that is possible when using two fingertips (pattern M3) is between 17.7 and 34.4% depending on the type of activities performed (rough manipulation, fine manipulation or manual exploration of the environment). It reaches 22.4 to 42.7% with three fingertips (pattern M4), 28.7 to 54.5% with four fingertips (pattern M5) and 33.3 to 61.7% with five fingertips (pattern M6). In theory, the more fingers you have the higher score you reach.

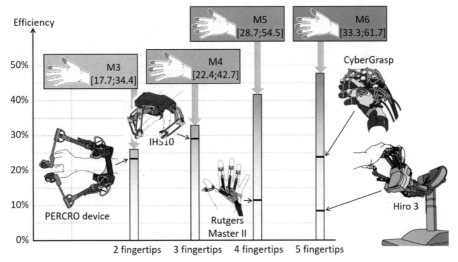

Fig. 8. Comparative study of the interaction efficiency of some existing fingertip input devices (figure adapted from [29]).

This is however not necessarily the case in practice as increasing the number of fingers the input device can track and apply force feedback on requires a higher number of DoFs, more links, joints and actuators. This added complexity tends to reduce the range of motion of the fingers, limit the force and stiffness available on each finger, increase friction and inertia, and globally limit the real efficiency. It appears indeed that, when taking into account all those criteria, the three fingers IHS10 glove [30] is more efficient than both the two fingers PERCRO device [31] and the four fingers Rutgers Master II [32] or the five fingers Cybergrasp [33] and Hiro 3 [34]. As a consequence, we will make use here of a three fingers device. In practice, this glove will be mounted at the tip of a Virtuose 6D and attached to the palm in order to allow for haptic feedback on both the palm and fingertips (this solution also allows compensating the glove's weight which has not to be carried on the hand).

2/Universal Fit: as explained in [35], two types of dexterous interfaces can be found in the literature. Exoskeletons, which are attached to every phalanges on which they can independently apply forces thanks to links and joints similar to the hand, allow simulating both precision and power grasps. They impose however hard mechanical constraints as their joints have to be roughly aligned with the fingers' ones. Hence, they must be tuned to each user, which is not convenient for a universal device that can be used by different operators. On the contrary, fingertip interfaces are fixed only on the palm and distal phalanges, and their geometry is less restricted, making them easily usable by different users. Their design is also much simpler. These advantages led us to focus on fingertip devices. To allow for natural interactions with the palm and fingers, links and joints have to be positioned and dimensioned so that the robot does not limit the fingers' movements.

3/High Transparency and Force Feedback Quality: haptic interfaces should be transparent in free space, i.e. display a mechanical impedance that is sufficiently low for the user to forget their presence. They should also be able to provide high impedances to simulate realistic contacts with stiff surfaces. This contradiction usually leads to a compromise between a high transparency in free space (i.e. low friction and inertia) and realistic force feedback in contact (i.e. high forces and stiffness). Here, we will exploit the fact that the slave hand has only one actuator per finger to control it with a glove having also only one actuator per finger, allowing to greatly simplify the design and favor a high transparency.

4.2 Electro-mechanical Design

The tri-digital master glove is illustrated in Fig. 9. It is composed of a base plate fixed on the palm and 3 robots allowing to track and apply forces on the distal phalanx of the thumb, index and middle finger to which they are attached. The base is dimensioned so that the index and middle robots' abduction-adduction axes best align with the fingers'

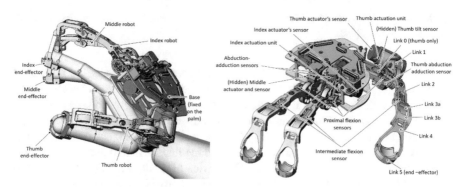

Fig. 9. Overview and main components of the tri-digital master glove (figure adapted from [8]).

ones (they are theoretically aligned for an adult man corresponding to the 50th percentile of the population).

Each robot is composed of 6 links (7 for the thumb, owing the requirement to make thumb opposition), allowing to move the fingers freely in their entire workspace. Joint sensors are integrated in the abduction-adduction, proximal flexion and intermediate flexion axes, as well as on the tilt axis of the thumb, allowing to compute the end-effectors' positions in space. Each robot is provided with a single actuation unit enabling force feedback at the fingertips roughly normal to the finger pulp (see Fig. 10). These actuators are equipped with high resolution incremental encoders, ensuring high quality position control. The design of the base plate, links dimensions and joints' range of motion were optimized in order to allow free movements of the fingers over their entire workspace. It is worth mentioning that, unlike gloves and exoskeletons whose dimensions fit specific users, fingertip devices can accommodate different hand sizes. Our device, whose design was inspired by the dexterous interface with hybrid haptic feedback for Virtual Reality applications presented in [35], can therefore easily be used by various users.

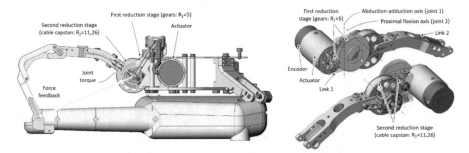

Fig. 10. Force feedback generated on fingertips and actuation units (figure adapted from [8]).

As shown in Fig. 10, force feedback is obtained with a Maxon REmax21 221028 DC motor (12 V, continuous torque 6.07 mNm, peak torque 17.3 mNm [21]) associated with a two stages reducer combining:

- A gear reducer with Delrin gears of 0.5 modulus allowing to obtain a reduction ratio of 5 (10 teeth primary gear HPC ZG0.5–10 glued on the motor axis, 50 teeth secondary gear HPC ZG0.5–50 as output).
- A miniature cable capstan reducer making use of a Berkley Whiplash Pro 0.42 mm Dyneema cable attached to pulleys of diameter 2.3 mm and 25.9 mm, hence a ratio of 11.26.

Such combination is highly transparent and backdriveable, yet compact and light. It ensures that, even if backlash occurs in the gear reducer, its amplitude is downscaled at the output of the cable capstan reducer, making it almost negligible in practice. It allows generating a continuous joint torque equal to 0.342 Nm and a peak joint torque of 0.974 Nm on the proximal flexion joint. This torque generates a force on the distal phalanx whose amplitude and direction depend on the finger configuration. Force is almost normal to the pulp when the finger is straight. The distance between the actuated axis and the fingertip being about 78.8 mm in this configuration for an adult man of medium size, continuous force is equal to 4.3 N and peak force to 12.4 N.

The motors are equipped with 512 ppt (2048 ppt after interpolation) magneto-optical encoders (ref. Maxon MR 201940). 1024 ppt Hall effect sensors are added at the joint level on the abduction-adduction axis and on the proximal and intermediate flexion axes (ref. Sensors RLS RM08 VB 00 10 B02 L2 G00, ref. Magnets RMM44 A3 A00). One can notice that the measurement of the proximal flexion is redundant. It is worth noting that both sensors are however not used for the same purposes.

- Owing the reduction ratio, the motor encoders give a very precise information, and they are co-located with the actuators. They are used for the position and force control (master and slave hands are linked using a bilateral position coupling scheme). However, these sensors do not allow to know the system configuration at start-up (these sensors are incremental).
- The role of the joint sensors is precisely to give an absolute joint angle value, avoiding the need for initialization when the glove is turned on.

4.3 Kinematics

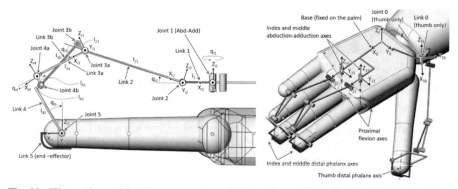

Fig. 11. Kinematic model of the master glove. A frame $R_{ik} = (O_{ik}, X_{ik}, Y_{ik}, Z_{ik})$ is associated with each link k of finger i, with $i = 1$ for the thumb, $i = 2$ for the index and $i = 3$ for the middle finger, with its origin positioned on the joint axis. q_{ik} is the rotation around joint k of finger i, and l_k (resp. l_{k1}, l_{k2}, l_{k3}) designates the length of link k (resp. of different parts of link k). An additional frame $R_{1b} = (O_{1b}, X_{1b}, Y_{1b}, Z_{1b})$ is introduced for the thumb which has an extra joint (figure adapted from [8]).

The simplified kinematic model of the master glove is illustrated in Fig. 11. Link 1 allows abduction-adduction while the other links allow finger flexion-extension. The links 2, 3a, 3b and 4 form an inverted parallelogram which allows the robot to remain close to the finger in its entire workspace. A pivot joint is added at the end of this structure to allow for the fingertip to rotate freely when the operator closes the hand.

It is worth noting that, contrary to the slave hand which comes from the field of robotics and for which we use Denavit Hartenberg (DH) conventions, the master hand device's design is inspired from the haptic glove presented in [35] which is used in Virtual Reality applications in which DH parameters are rarely used. In practice, we use the more simple notations given in Fig. 11, with which the kinematic model of the thumb, index and middle robots can be written as follows (with $i = 1$ for the thumb, $i = 2$ for the index and $i = 3$ for the middle finger):

$$T_{i\,01} = \text{trans}(X_0,\ d_{ix}).\text{trans}(Y_0,\ d_{iy}).\text{rot}(Z_0,\ q_{i1}) \tag{5}$$

$$T_{i\,12} = \text{trans}(X_{i1},\ l_1).\text{rot}(Y_{i1},\ q_{i2}) \tag{6}$$

$$T_{i\,23} = \text{trans}(X_{i2},\ l_{21}).\text{rot}(Y_{i2},\ q_{i3}) \tag{7}$$

$$T_{i\,34} = \text{trans}(X_{i3},\ l_{3a}).\text{rot}(Y_{i3},\ q_{i4}) \tag{8}$$

$$T_{i\,45} = \text{trans}(Z_{i4},\ -l_{41} - l_{43}).\text{trans}(X_{i4},\ l_{42}).\text{rot}(Y_{i4},\ q_{i5}) \tag{9}$$

For the thumb, Eq. (5) is replaced with the following equations:

$$T_{1\,b0} = \text{trans}(X_0,\ d_{1x}).\text{trans}(Y_0,\ d_{1y}).\text{trans}(Z_0,\ d_{1z}).\text{rot}(Z_0,\ q_{1z0}).\text{rot}(X_{1b},\ q_{1x0}) \tag{10}$$

$$T_{1\,01} = \text{trans}(Z_{10},\ l_0).\text{rot}(Z_{10},\ q_{11}) \tag{11}$$

In these equations, q_{i1}, q_{i2} and q_{i3} are given by the motors' and joints' sensors, while q_{i4} can be computed from q_{i3} by solving the equations of the inverted parallelogram as explained in [35] and [36]. The distal rotation q_{i5} is not measured nor computed as the position of the fingertip, which is the information of interest, does not depend on it.

4.4 Manufacturing and Integration

The manufactured prototype made of aluminum parts and its main components are shown in Fig. 12. A specific attention was given to the routing of the actuators and sensors cables which are guided along the robots' structure so that they pass the closest possible to the joints' axes in order to resist as less as possible to the links' movements. A custom designed PCB, integrated in the base plate, is used to connect the glove to its controller. This PCB is in charge of both powering the sensors and actuators and of conditioning and filtering the sensors' signals. In order to favor modularity, each finger is connected to this PCB through a specific connector on the robots' side, and three connectors are used on the controller side, respectively in charge of the actuators power supply, motors' encoders and joint sensors. As shown in Fig. 13, this PCB is protected by a thin plastic part, and the base is attached to a mitten through a 3D printed part. This way, the glove can be easily put on or taken off.

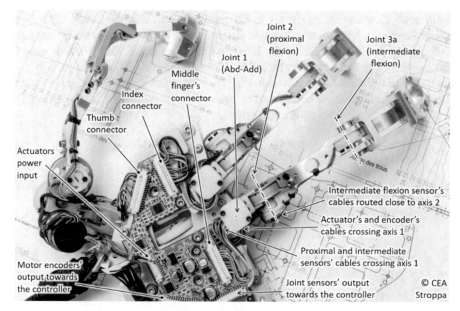

Fig. 12. Main components and cabling of the manufactured dexterous hand master prototype.

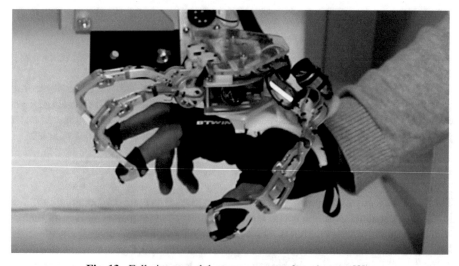

Fig. 13. Fully integrated dexterous master glove (source [8]).

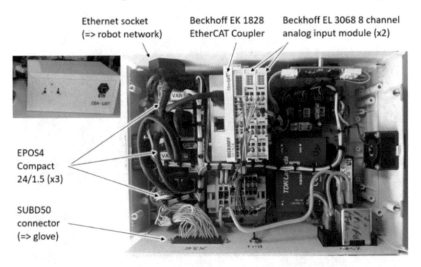

Fig. 14. Dexterous hand master controller (source [8]).

The haptic master glove is controlled using an Ethercat controller illustrated in Fig. 14. Three Maxon EPOS4 Compact 24/1.5 modules are used for controlling the actuators, while a Beckhoff EK1828 Ethercat Coupler connected to two Beckhoff EL 3068 analog input modules with 8 channels each (0–10 V, 12 bits) is used to connect the joint sensors. The controller also integrates 12 V (for the EPOS4 modules) and 24 V (for the Beckhoff modules and the glove PCB) power supplies. It is connected to the glove through a SUBD 50 connector and to the robot through an Ethernet socket.

5 Bi-manual Teleoperation Controller

5.1 Detailed Presentation of the Bi-manual Teleoperation Set-Up

As previously stated, our bi-manual teleoperation platform is composed of two robots on the slave side, a Stäubli RX160 equipped with the three fingers gripper and a prototype Isybot cobot (https://www.isybot.com/) provided with a bi-digital gripper with parallel jaws both developed at CEA LIST (see Fig. 15). This configuration allows efficiently performing dexterous operations with both hand, as human naturally do in everyday life, either to move an object with both hands or to hold it with one hand while making an operation on it with the other hand.

All CEA devices (i.e. all but the RX160) make use of ball screw reducers. They come with direct transmissions on the three digital gripper as presented above and on the second robot pair of pliers with parallel jaws whose design will not be detailed here, and with a secondary cable transmission as presented in [22] on the prototype cobot whose design will neither be presented here. This particular type of reducers and transmissions are key innovative elements enabling these robot and grippers to have a very high level of mechanical transparency (i.e. very low friction and inertia) and back-drivability. This solution provides accurate joint torque estimates from simple current sensing, without the

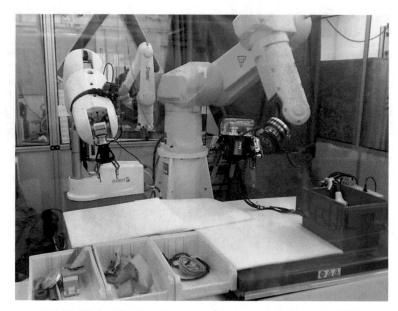

Fig. 15. Slave side of the bi-manual dexterous teleoperation platform.

need for delicate electronic sensors which would fail in high beta and gamma radiation environments as encountered on nuclear sites. Note that current sensing is also used in conventional robot arms to provide torque estimates, but these measurements are significantly erroneous due to the higher inertia and friction of conventional actuators. As a consequence, the RX160 is equipped with an ATI 6 DoF force/torque sensor placed

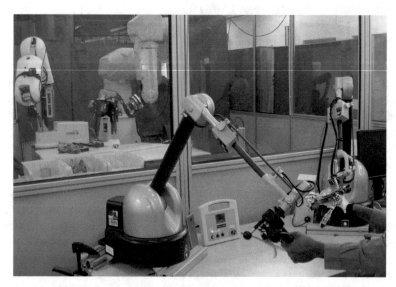

Fig. 16. Master side of the bi-manual dexterous teleoperation platform.

between its end-effector and the gripper, allowing a direct measure of the efforts applied on the robot. When moving to a real environment, this sensor will have to be replaced with a radiation hardened sensor with a remote electronics as in [3].

As shown in Fig. 16, these robots are controlled using two input devices mapping the slave-arm capabilities. A Virtuose 6D TAO master arm from Haption is used to control the Isybot prototype with 6DoF force feedback. This input device is further equipped with an instrumented handle which enables a fine control of the gripper's closing and opening movement. A second Virtuose 6D TAO is used to control the Stäubli RX160. It is worth noting that the Virtuose is provided with a universal handle adapter, allowing to easily shift from one handle to another. The original handle is replaced here with our novel tri-digital master glove, allowing fine control of the dexterous gripper with force feedback on both the palm and fingers. This set-up is intended to enable high fidelity dexterous force-reflection tele-presence.

5.2 Master-Slave Coupling

Our bi-manual teleoperation platform is controlled using the TAO framework. TAO is a teleoperation middleware allowing a high-speed synchronization and control of several real or virtual mechanisms (e.g. master arms, slave arms, dynamic simulators) [4]. Among the different TAO teleoperation modes, we make use here of master-slave bilateral position coupling schemes as they provide a compelling force feedback yet prove to be passive (i.e. stable whatever the actions of the human operator and the efforts applied on the slave robots) when the control gains are properly tuned.

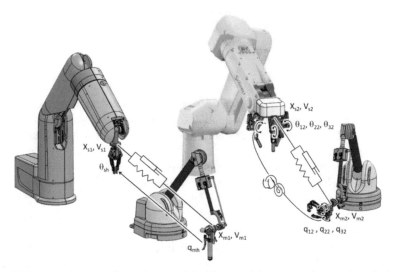

Fig. 17. Master-slave coupling schemes of the bi-manual dexterous teleoperation platform.

As shown in Fig. 17, different implementations of this bilateral position couplings are used in practice:

- The Isybot prototype and the associated Virtuose 6D master arm, as well as the Stäubli RX160 and the second Virtuose 6D, are coupled in the Cartesian space as they have different kinematics. Denoting X_{s1} and X_{s2}, respectively X_{m1} and X_{m2}, the configuration of their end-effector, respectively of their handle, computed from their joint measurements using their direct geometric model, and V_{s1}, V_{s2}, V_{m1} and V_{m2} their kinematic twists evaluated with their Jacobian matrix $J_{si}(\theta_i)$ (respectively $J_{mi}(q_i)$) expressed in the base frame and reduced to the center of the end-effector, we can write these coupling schemes as follows:

$$F_{si} = -H_f.F_{mi} = K.\left(X_{mi} - H_p.X_{si}\right) + B.\left(V_{mi} - H_p.V_{si}\right) \qquad (12)$$

With: K the position control gain expressed in N/m and Nm/rad (K is usually represented as a spring, as shown in Fig. 17).
B the speed gain expressed in N/m/s and Nm/rad/s (B is usually represented as a damper).
H_p an optional position scaling factor ensuring a multiplied movement on the slave side compared to the master side (the master movements are H_p times larger than the slave ones).
H_f an optional force and torque scaling factor allowing the slave arms to produce efforts H_f times larger than the forces applied on the master arm.

The torques applied on the different joints of the master and slave arms to ensure these coupling are then computed as follows:

$$\tau_{mi} = J_{mi}^T(q_i).F_{mi} \qquad (13)$$

$$\tau_{si} = J_{si}^T(\theta_i).F_{si} \qquad (14)$$

It is worth noting that the scaling factors are very convenient when the master and slave arms have different workspaces and/or force and torque capacities. They should however be used with care as they introduce several changes in the stimuli perceived by the human operator.

First, a large position scaling factor allows large slave displacements with limited hand movements, hence reducing fatigue, but small and precise slave movements become very difficult as every single small hand displacement is amplified and produces a large slave movement. Fortunately, the TAO framework also implements a software clutch function, i.e. the master and slave arm are coupled only if the clutch is engaged, and they can move independently when it is disengaged. In practice, it is possible to control the slave robots by moving the master arms as long as the handle remain in an ergonomic and comfortable volume, and to disengage the clutch as soon as their configuration is no more convenient, then to move back the master arms in more comfortable configurations and re-engage the clutch to continue working. This way, the slave robots can span a much larger volume than the master arms without the need for large position scaling factors. In practice, the position scaling factor were kept equal to 1 here.

The same is true for the force scaling factors. Large scaling reduces fatigue but diminishes the user sensitivity. Also, if the position and force scaling factors differ, the perceived environment dynamics will be modified (e.g. with a position factor of 1, i.e. the master and slave move the same amplitude, and a force factor of 2, i.e. the slave applies two times more force than the master, the user will perceive the contacted objects as two times softer as half the same force will produce the same movement). In practice, force scaling factors were kept low, i.e. equal to 1.

- As the three fingers slave hand and the tri-digital hand master are underactuated, it is not possible to couple them in Cartesian space. They are coupled in joint space instead and their behavior is the governed by the following equation:

$$\tau_{si2} = -h_f.\tau_{mi2} = k.(q_{i2} - h_p.\theta_{i2}) + b.(\dot{q}_{i2} - h_p.\dot{\theta}_{i2}) \qquad (15)$$

With: τ_{si2} (respectively τ_{mi2}) the joint torques applied on the second axis of the finger i of the slave hand (respectively the master hand device).
k the joint position control gain expressed in Nm/rad.
b the joint speed gain expressed in Nm/rad/s.
h_p an optional joint position scaling factor.
h_f an optional joint torque scaling factor.
q_{i2} (respectively $\dot{q}_{i2}$) the joint position (respectively the joint speed) of the second joint (i.e. the actuated joint) of the ith finger of the tri-digital master hand device (see Fig. 11).
θ_{i2} (respectively $\dot{\theta}_{i2}$) the joint position (respectively the joint speed) of the second joint (i.e. the actuated joint) of the ith finger of the three fingers slave hand (see Fig. 3).

In practice, the position scaling factors were tuned so that the slave hand closes completely when moving the master hand. The force required to grasp the large and heavy objects encountered in our use-case being much higher than the human hand grasping forces, themselves higher than the master hand torques, a 1/10 force scaling factor (i.e. the maximum value allowed by TAO) was used to couple the master and slave hands.

- As the instrumented handle of the Virtuose 6D is not actuated, it is not possible to provide force feedback on the left hand grasping movement and the bi-digital gripper's opening movement θ_{sh} is controlled in position in open loop so as to follow the handle movement q_{mh}.

5.3 Tri-Digital Master-Three Fingers Slave Hand Mapping

As previously stated, the master hand-slave hand coupling is performed using a bilateral position coupling scheme implemented in the joint space. This solution was chosen to keep the coupling as intuitive as possible for the user, so as to limit his or her cognitive load.

In practice however, it is difficult to control the slave hand intuitively if the master fingers are always coupled with the same slave fingers as the slave hand can have configurations that largely differ from the human hand. If for example we couple the thumb with the fixed slave finger and the index and middle with the moving fingers, the slave hand behavior will be natural for cylindrical grasps as the slave 'thumb' will face the other two fingers. The spherical grasps will also be quite natural as the slave fingers will encompass the object just as the user fingers would do in direct manipulation. The planar grasps will however not be natural at all, as the fingers controlling the slave hand's closing and opening movement will be the index and middle fingers while one would expect planar grasps to occur between the thumb and the index or middle. And the same limitations appear whatever the choice of fixed master-slave fingers association. To cope with this issue, we developed an adaptive mapping strategy allowing to keep natural master hand glove configurations whatever the type of grasp performed by the slave hand. With respect to Fig. 18, the principle is the following:

- In planar grasps, only the moving fingers of the slave hand (i.e. fingers A and C) are active and involved in the coupling (finger B is blocked in open position). The thumb and index are coupled with fingers A and C (the middle is not used).
- In spherical grasps, the thumb is coupled with finger A, the index with finger B and the middle finger with finger C.
- Finally, in cylindrical grasps, finger A and C have to move in the same manner in order to ensure a correct closure. To meet this requirement, the thumb is coupled with finger A and the index with finger B. Then finger C is coupled with finger A, and the middle is coupled to finger C. This way, fingers A and C move similarly, as well as the thumb and the middle fingers.

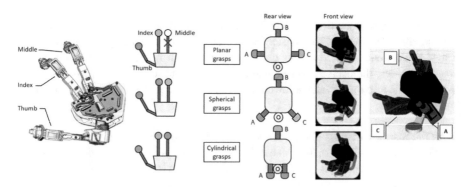

Fig. 18. Master and slave fingers mapping.

In practice, the change from one grasp to another is managed by a state machine. It was first tested to use the master thumb abduction-adduction imposed by the operator's hand to choose the grasp type and then to maintain it until the object is fully grasped. It appeared however that using a simple external selector is more comfortable to switch the abduction configuration between three main positions.

First tests consisted in verifying that the slave hand can be controlled by moving the master glove, and reversely that the master hand reproduces the slave hand's motions (see Fig. 19). Then the master glove was used to remotely grasp various object with force feedback (see Fig. 20).

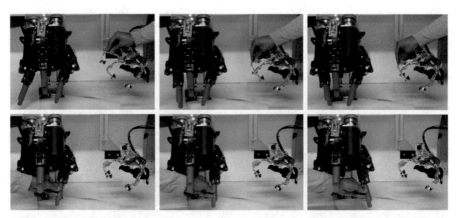

Fig. 19. Functional validation of the bilateral joint position coupling scheme. Top: direct sense (from the master middle to the slave). Bottom: inverse sense (from the slave to the master thumb).

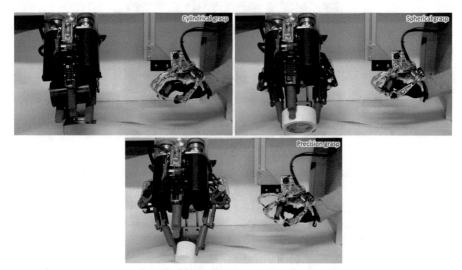

Fig. 20. Functional validation of teleoperated cylindrical, spherical and planar grasps (adapted from [8]).

6   Bi-manual Teleoperation Tests and Validation

Once validated, the master hand glove was mounted on a Virtuose 6D master arm, and it was further associated with a second master slave system. The resulting bi-manual teleoperation setup was used to test the ability of an operator to grasp and manipulate several types of objects similar to the waste found in nuclear containers (e.g. piece of cloth, rigid objects, cables, etc.). As shown in Fig. 21, these operations were successful. It was even possible to pass objects from one robot to the other.

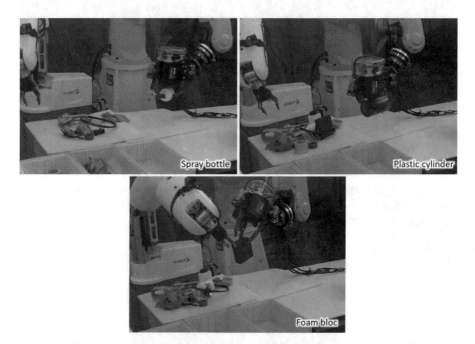

Fig. 21. Grasping various objects in bi-manual teleoperation (adapted from [8]).

7   Conclusions and Perspectives

This paper presents a novel bi-manual dexterous teleoperation setup for nuclear waste remote manipulation. This platform is composed of an underactuated three fingers slave gripper mounted on a large capacity slave robot and a simple bi-digital gripper with parallel jaws fixed on a cobot prototype. These slave robots are controlled using two Virtuose 6D TAO master arms, one of which is equipped with a novel tri-digital master hand device. Master and slave robots are coupled using a bilateral position coupling scheme implemented in the Cartesian space for the robots and in the joint space for the hands, with a mapping strategy allowing to intuitively control cylindrical, spherical and planar grasps. This system proves to be able to grasp various objects similar in size and shape as those encountered in waste containers. The operator is able to extract the objects, pass them from one robot to the other and sort them in various deposit boxes. This functional tests validate the ability of this system to perform nuclear sort and segregation tasks. Future work should be dedicated to a more precise experimental characterization of the platform performances.

Acknowledgments. The authors would like to thank the colleagues who contributed to the technical developments presented in this article, especially D. Diallo, B. Perochon and P. Chambaud. This research was partly supported by the Horizon 2020 RoMaNS project (Robotic Manipulation for Nuclear Sort and Segregation, #645582) funded by the European Commission.

References

1. Köhler, G.W.: Typenbuch der Manipulatoren - Manipulator type book. Verlag Karl Thiemig, München, Germany, Thiemig Taschenbücher (1981)
2. Vertut, J., Coiffet, P.: Les robots - Tome 3A : téléopération, évolution des technologies. Hermès Publishing, Paris, France (1984)
3. Piolain, G., Geffard, F., Coudray, A., Garrec, P., Thro, J.F., Perrot, Y.: Dedicated and standard industrial robots used as force-feedback telemaintenance remote devices at the AREVA recycling plant. In: Proceedings 1st IEEE International Conference on Applied Robotics for the Power Industry, Montreal, QC, Canada, pp. 1–6 (2010)
4. Geffard, F., et al.: TAO2000 V2 computer-assisted force feedback telemanipulators used as maintenance and production tools at the AREVA NC-la hague fuel recycling plant. J. Field Rob. **29**(1), 161–174 (2012)
5. European Commission Portal, Robotic Manipulation for Nuclear Sort and Segregation (ROMANS). https://cordis.europa.eu/project/id/645582. Accessed 02 Mar 2022
6. World Nuclear News, Funding for waste-handling robotics development. https://www.world-nuclear-news.org/NN-Funding-for-waste-handling-robotics-development-2502154.html. Accessed 02 Mar 2022
7. Marturi, N., et al.: Towards advanced robotic manipulation for nuclear decommissioning: a pilot study on tele-operation and autonomy. In: Proceeding of International Conference on Robotics and Automation, Humanitarian Applications, Amritapuri, India, p. 8 (2016)
8. Gosselin, F., Grossard, M., Diallo, D., Perochon, B., Chambaud, P.: Design and development of a dexterous master glove for nuclear waste telemanipulation. In: Proceedings 18th International Conference on Informatics in Control, Automation and Robotics, pp. 459–468 (2021)
9. Kochan, A.: Shadow delivers first hand. Ind. Robot Int. J. **32**(21), 15–16 (2005)
10. Walker, R., De La Rosa, A., Elias, H., Godden, M., Goldsmith, J.: Advances in Actuation Technology for Compliant Dextrous Manipulation. In: Proc. IEEE Int. Conf. on Robotics and Biomimetics, pp. 1429–1433, Tianjin, China (2010)
11. Martin Amezaga, J., Grossard, M.: Design of a fully modular and backdriveable dexterous hand. Int. J. Rob. Res. **33**(5), 783–798 (2014)
12. Grebenstein, M., et al.: The DLR hand arm system. In: Proceedings of IEEE International Conference on Robotics and Automation, Shanghai, China, pp. 3175–3182 (2011)
13. Cutkosky, M.R.: On grasp choice, grasp models, and the design of hands for manufacturing tasks. IEEE Trans. Rob. Autom. **5**(3), 269–279 (1989)
14. Feix, T., Pawlik, R., Schmiedmayer, H.B., Romero, J., Kragic, D.: A comprehensive grasp taxonomy. In: Robotics, Science and Systems Conference Workshop on Understanding Human Hand for Advancing Robotic Manipulation, Seattle, WA, USA, pp. 58–59 (2009)

15. Fishel, J.A., et al.: Tactile telerobots for dull, dirty, dangerous, and inaccessible tasks. In: Proceedings of IEEE International Conference on Robotics and Automation, Paris, France, pp. 11305–11310 (2020)
16. Mnyusiwalla, H., Vulliez, P., Gazeau, J.P., Zeghloul, S.: A new dexterous hand based on bio-inspired finger design for inside-hand manipulation. IEEE Trans. Syst. Man Cybern. **46**(6), 809–817 (2016)
17. Yuan, S., Epps, A.D., Nowak, J.B., Salisbury, J.K.: Design of a roller-based dexterous hand for object grasping and within-hand manipulation. In: Proceedings of IEEE International Conference on Robotics and Automation, Paris, France, pp. 8870–8876 (2020)
18. Townsend, W.: The BarrettHand grasper - programmably flexible part handling and assembly. Ind. Robot: Int. J. **27**(3), 181–188 (2000)
19. Birglen, L., Gosselin, C.M.: Kinetostatic analysis of underactuated fingers. IEEE Trans. Rob. Autom. **20**(2), 211–221 (2004)
20. Birglen, L., Laliberté, T., Gosselin, C.: Underactuated Robotic Hands. Springer Tracts in Advanced Robotics, vol. 40. Springer, Heidelberg (2008). https://doi.org/10.1007/978-3-540-77459-4
21. Maxon : Maxon motor catalogue, programme 2016–2017 (2016)
22. Garrec, P.: Screw and cable actutators (SCS) and their applications to force feedback teleoperation, exoskeleton and anthropomorphic robotics. In: Robotics 2010: Current and Future Challenges, Intech: Rijeka, Croatia, pp. 167–191 (2010)
23. Khalil, W., Dombre, E.: Modélisation, identification et commande des robots, 2nd edn. Hermès Science Publications, Paris (1999)
24. Bogue, R.: Exoskeletons and robotic prosthetics: a review of recent developments. Ind. Robot: Int. J. **36**(5), 421–427 (2009)
25. Foumashi, M.M., Troncossi, M., Parenti Castelli, V.: State of the-art of hand exoskeleton systems. Internal report, Univ. Bologna, p. 54 (2011)
26. Heo, P., Min, G.G., Lee, S.J., Rhee, K., Kim, J.: Current hand exoskeleton technologies for rehabilitation and assistive engineering. Int. J. Prec. Eng. Manuf. **13**(5), 807–824 (2012)
27. Gopura, R.A.R.C., Bandara, D.S.V., Kiguchi, K., Mann, G.K.I.: Developments in hardware systems of active upper-limb exoskeleton robots: a review. Rob. Auton. Syst. **75**(B), 203–220 (2016)
28. Perret, J., Van der Poorten, E.: Touching virtual reality: a review of haptic gloves. In: Proceedings of 16th International Conference on New Actuators, Bremen, Germany, pp. 270–274 (2018)
29. Gonzalez, F., Gosselin, F., Bachta, W.: Analysis of hand contact areas and interaction capabilities during manipulation and exploration. IEEE Trans. Haptics **7**(4), 415–429 (2014)
30. Gosselin, F.: Guidelines for the design of multi-finger haptic interfaces for the hand. In: Proceedings of 19th CISM-IFToMM Symposium on Robot Design, Dynamics and Control, Paris, France, pp. 167–174 (2012)
31. Frisoli, A., Simoncini, F., Bergamasco, M., Salsedo, F.: Kinematic design of a two contact points haptic interface for the thumb and index fingers of the hand. ASME J. Mech. Des. **129**(5), 520–529 (2007)
32. Bouzit, M., Burdea, G., Popescu, G., Boian, R.: The Rutgers master II—New design force-feedback glove. IEEE/ASME Trans. Mechatron. **7**(2), 256–263 (2002)
33. Aiple, M., Schiele, A.: Pushing the limits of the CyberGraspTM for haptic rendering. In: Proceedings of IEEE International Conference on Robotics & Automation, Karlsruhe, Germany, pp. 3541–3546 (2013)
34. Endo, T., et al.: Five-fingered haptic interface robot: HIRO III. IEEE Trans. Haptics **4**(1), 14–27 (2011)

35. Gosselin, F., Andriot, C., Keith, F., Louveau, F., Briantais, G., Chambaud, P.: Design and integration of a dexterous interface with hybrid haptic feedback. In: Proceedings of 17th International Conference on Informatics in Control, Automation and Robotics, pp. 455–463 (2020)
36. Ngalé Haulin, E., Lakis, A.A., Vinet, R.: Optimal synthesis of a planar four-link mechanism used in a hand prosthesis. Mech. Mach. Theory **36**(11–12), 1203–1214 (2001)

Pose Optimization of Task-Redundant Robots in Second-Order Rest-to-Rest Motion with Cascaded Dynamic Programming and Nullspace Projection

Moritz Schappler[(✉)] [iD]

Institute of Mechatronic Systems, Leibniz University Hannover,
An der Universität 1, 30823 Garbsen, Germany
`moritz.schappler@imes.uni-hannover.de`

Abstract. An optimal trajectory for the redundant coordinate for robots in tasks with rotational symmetry such as machining has to be found to ensure good performance and overall feasibility. Due to high nonlinearity of performance criteria especially for parallel robots a sole local optimization may lead to infeasible solutions for large-scale motion. A pointing task consisting of multiple rest-to-rest trajectories with given dense sample times is regarded as given. Constraints regarding system limits on position, velocity and acceleration have to be met. The proposed algorithm combines nullspace projection for local optimization between the rest poses with dynamic programming at the rest poses in a cascaded scheme to optimize the rotation around the tool axis. Applications to other types of redundancy are also possible. The proposed local/global optimization scheme only needs wide discretization of the redundant coordinate and therefore has acceptable computational performance for offline optimization of robot motion. It is able to find feasible and near-optimal trajectories for a six-degree-of-freedom (DoF) parallel robot in several exemplary five-DoF tractories with many constraints.

Keywords: Robot manipulator · Task redundancy · Inverse kinematics · Trajectory optimization · Dynamic programming · Nullspace projection

1 Introduction and State of the Art

Resolution of kinematic redundancy is a persistent topic in robotics research. Task redundancy or functional redundancy is a special case if the task requires less DoF than are controlled by the end effector in it's operational space regardless of the dimension of the joint space. It is explicitly relevant for (fully) parallel

© The Author(s), under exclusive license to Springer Nature Switzerland AG 2023
O. Gusikhin et al. (Eds.): ICINCO 2021, LNEE 1006, pp. 106–131, 2023.
https://doi.org/10.1007/978-3-031-26474-0_6

robots where not the joint space as for serial robots, but the operational space of the moving platform is the essential structural kinematic characteristic. The focus on the following discussion of related work and of the paper's examples is therefore put on parallel robots. However, the proposed approach of this paper is not restricted to task redundancy or parallel robots and can be directly transferred to serial robots with one DoF of redundancy.

1.1   State of the Art and Related Work

Kinematic redundancy can be distinct in intrinsic redundancy, (e.g. 7-DoF serial robots in 6-DoF tasks), and task redundancy, as e.g. for 6-DoF robots in 5-DoF tasks. The degree of intrinsic [task] redundancy is the excess of dimension of the joint space [operational space] over the task space. A formal definition of redundancy is given in [11,15,29] for serial robots and in [8] for parallel robots.

Applications for task redundancy are mainly tasks with rotational symmetry of the tool or the process in general. Machining tasks using serial robots are discussed for milling in [18,19,34] or for drilling in [35]. Other examples using serial robots are arc welding [11] and fiber placement for composite materials [5]. The main difference of these tasks from a kinematic point of view is that milling tasks (and arc welding) usually require redundancy optimization for a continuous trajectory (which is also the focus of this paper) and drilling (or spot welding) only requires an optimization for approaching the pointing pose. Examples for parallel robots' task redundancy with more than one redundant DoF are end milling [30] or milling with a spherical cutter [32], where all three rotational coordinates can be subject to optimization.

The *objective of exploiting the redundancy* is for once to improve feasibility by avoiding joint limits [11,35] or singularities. Obstacle collision avoidance can be implemented by minimizing corresponding potential functions [4,10,21]. Singularity avoidance can be implemented as an optimization objective e.g. by

- the joint space distance to the first joint configuration that violates a parameter of singularity [11],
- the squared condition number of the robot Jacobian [2,15,35],
- the condition number of the parallel robot forward kinematics Jacobian [7],
- a fraction containing all singular values of the Jacobian [26],
- the Jacobian's determinant [1]
- directly using the Jacobian's condition number [28].

The homogenized pose error can be minimized as a measure for accuracy as well as singularities [12], Other optimization criteria may be rather specific such as milling process stability [19]. In general, any optimization criterion dependent on the robot configuration can be used, with certain restrictions to continuous differentiability depending on the optimization method.

The potential function of the optimization objective can be visualized well over a trajectory for the case of *redundancy of degree one*. This *performance (criterion) map* was introduced by Wenger for serial robots and parallel robots,

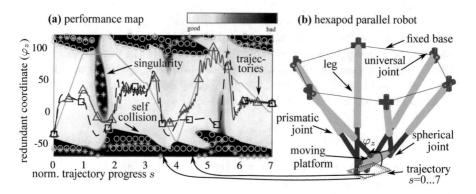

Fig. 1. Exemplary surface map of a performance criterion (**a**) for a parallel robot (**b**).

as summarized in [25], there termed "feasibility map". This map is especially useful for visual inspection in the case of large-scale motion relative to the robot workspace size. An exemplary map is given Fig. 1 for a hexapod robot with redundant orientation coordinate φ_z, which may be the irrelevant rotation around a tool axis. The map was e.g. used in [25] to create cycloidic trajectories directly in the performance map to plan changes between the 2^3 working modes (elbow configurations) of a planar parallel robot. A "robot transmission ratio map" was used in [34] to optimize the trajectory for a six-axis serial robot in a milling task. Maps called "state space grids" were used in [4] for a seven-DoF serial robot in six-DoF tasks. The maps for the 2^3 extended aspects (e.g. base, elbow and hand configurations) were created for one joint angle as the redundancy parameter.

For *two degrees of redundancy* the heatmap visualisation for the trajectory is not possible any more and a two-dimensional map can be given for single poses of the robot. This has been shown for seven-DoF robots in five-DoF tasks such as medical laser operations [27] or by using an additional rotary table for an industrial robot in five-DoF milling tasks [18]. The performance map is then build by discretization of one joint axis and the end effector rotation.

Different geometric methods for redundancy resolution are available, mainly with the goal to separate the kinematic formulation into task and redundant coordinates for the subsequent optimization. A non-exhaustive list comprises

- task space orthogonal decomposition [2,15] and twist decomposition [11],
- separation of joint coordinates in redundant and non-redundant on position level for serial [23] and parallel [16] robots or on velocity level [24],
- giving a parametric form of the dynamics equation with the redundant end effector coordinate [22],
- expressing the end effector angular velocity in the local frame and removing the last component corresponding to the redundant coordinate [24,36] (for serial robots), trivial for the planar case [1] (e.g. for a planar parallel robot),

– formulating the inverse kinematics on position level and exploiting nonlinear orientation to eliminate the redundant task space coordinate by using a Z-Y-X Tait-Bryan-angle residual [28] (for position level and higher order).

For online optimization and fast offline planning nullspace projection is the most efficient *optimization algorithm* for the differential inverse kinematics. It is the basis for many works on serial robots using locally optimal redundancy resolution [3, 11, 14, 21, 24, 36]. Due to more sophisticated modeling, *nullspace projection* is more scarce in parallel robot inverse kinematics literature. The method is used mainly for kinematic redundancy [26] and a review on "redundancy in parallel mechanisms" [8] does not mention the case of task redundancy at all. It is introduced e.g. in [1] for the trajectory of a planar parallel robot and in the author's previous work [28] generally for spatial (fully) parallel robots.

Most formulations presented above are on *velocity level (first order)* which hinders trajectory optimization using acceleration and actuator force constraints like in [26], which can be avoided by an *acceleration-level (second-order)* formulation. Using only proportional feedback of the optimization objective potential there as in [1, 21] may lead to unwanted oscillations, which might be encountered by a PD feedback [3, 28] or an analytic derivative of the nullspace projection [24].

Other approaches than nullspace projection for trajectory redundancy optimization can be chosen from various *existing optimization algorithms* such as

– SQP [15] (for an industrial robot, without input or state constraints),
– the Simplex algorithm and quadratic programming [22] (hexapod example),
– discrete optimization of a redundant coordinate at rest poses of a kinematically redundant planar parallel robot [12],
– interval analysis [16] (at the example of a hexapod robot),
– (binary) small-angle perturbation [6] of the redundant coordinate of a task-redundant planar parallel robot.
– bracketing and bisection, Brent-Dekker method [35],
– genetic algorithms (GA) [30] (for three redundant coordinates of a hexapod).

While apart from the heuristic GA, the aforementioned algorithms also are only locally optimal like nullspace projection, *dynamic programming* (DP) is a deterministic *global optimization* algorithm. It is based on discretization of the optimization variable (redundant coordinate) on discrete stages (time steps). Examples for using *dynamic programming in robotics* are

– optimizing the time evolution of a given trajectory [31] as an early work,
– using Pontryagin's maximum principle for kinematic redundancy [20],
– investigation of Pareto optimality for multiple objectives for pose optimization of a seven-DoF serial robot [9],
– position-level optimization of two redundant DoF of a seven-DoF robot [27],
– wide and fine discretization of redundant joint coordinates in a static pose optimization of an industrial robot with an additional positioner [5].

For continuous trajectory optimization the method of *differential dynamic programming* (DDP) is advantageous. An overview and an elaboration on using the method on problems consisting of multiple phases is given in [13]. Each rest-to-rest motion can be considered as such a phase and each time step as a stage. Several works have used DDP in robotics, such as

- motion planning of a humanoid robot and a contribution regarding implementing box constraints in DDP [33],
- a constraint approach by an augmented Lagrangian validated at an example of reaching a target position with a seven-DoF arm [10],
- an application of DDP to task-redundant parallel robots [26] in combination with nullspace projection on velocity level.

1.2   Summary of the State of the Art and Scenario of the Paper

The main requirements on the trajectory optimization algorithm developed in this paper are motivated by machining tasks and other processes, where task redundancy of degree one exists and the desired task space trajectory of the end effector is given, including the time profile. The task is considered to consist of connected second-order rest-to-rest trajectories e.g. from G-code in CNC machining, which ensures a continuous velocity profile. Extending the resting condition also to the nullspace motion has the potential to reduce oscillations and to decouple the optimization. A feasible and desirably optimal trajectory for the redundant coordinate has to be found. A kinematic formulation is sufficient but can be extended by dynamics constraints regarding actuator force limits.

For the sake of simplicity, the *reconfiguration of the robot* between working modes is out of scope of this paper assuming a trajectory that can be performed only with one working mode. One reason is that distinct reconfiguration motion [12] reduces productive times of the machine. Reconfiguration during robot motion as discussed in [25] for parallel robots requires flipping the elbow configuration of a passive joint, which might not be possible to detect by sensors. Further, this motion leads through kinematic singularities, which can have unwanted side-effects. Additionally, the complete discretization of a performance map for all working modes of the robot is necessary for each task space trajectory. Ideally, the computation of the performance map is no requirement for obtaining the full trajectory and is only used for optional visualization.

As stated in [13], usually DDP approaches consider only one phase, i.e. one robot rest-to-rest motion. Combining rest-to-rest trajectories presents a special case regarding optimization, which has high practical relevance. DDP requires computing the variational formulation of the optimization problem with Jacobian and eventually Hessian w.r.t. input and state variables. This can be omitted by using just local optimization which requires only robot Jacobians and gradients of their performance criteria which are easy to obtain. Performing an unknown number of iterations of forward and backward passes in DDP can be avoided by using DP. In summary a trajectory optimization algorithm below the complexity of DDP for the envisioned use case scenario would increase robot performance in the given task.

1.3 Contribution of the Paper

The optimization of multiple rest-to-rest (i.e. multi-phase) trajectories can be solved by connecting the *local optimality of nullspace projection* with the *global optimality of dynamic programming*. This allows to use the local nullspace projection scheme for the optimization of the redundant coordinate between rest poses in a fine sampling time with continuous states. The approximation of global optimality of the trajectory is achieved with dynamic programming of the same optimization variable at rest poses based on the actual obtained coordinates from the local optimization. Thereby an *extension to dynamic programming* is proposed such that state values of the optimization variable are not given as discrete values, but as an *interval* determined by the final value of the nullspace projection. The performance of this approximation depends on the quality of the local optimization and the underlying assumption of limited changes of the feasibility map during one phase.

The contributions of the paper therefore are in summary

- a cascaded scheme of dynamic programming and nullspace optimization,
- an extension to dynamic programming regarding interval optimization,
- the application of the scheme on second-order rest-to-rest robot motion,
- the extension of the second-order nullspace projection inverse kinematics approach presented in [28] to support the cascaded dynamic programming.

The problem from the author's preceding conference paper [28] is solved globally using the new dynamic programming scheme. The remainder of the paper is structured as follows. The problem description in the sense of optimization is presented in Sect. 2. The nullspace optimization scheme for the robot from [28] is revised and extended in Sect. 3. Finally, the methods are combined for trajectory optimization in Sect. 4 with simulation results shown in Sect. 5.

2 Problem Description: Task Redundancy Optimization

A robot manipulator, either serial or parallel, is able to create motion with it's end effector in the operational space. The corresponding end effector pose variable is denoted as $\boldsymbol{x}_E = (r_x, r_y, r_z, \varphi_x, \varphi_y, \varphi_z)^{\mathrm{T}}$ and contains position $\boldsymbol{r}$ and orientation $\boldsymbol{\varphi}$. The latter is expressed with the $X\text{-}Y'\text{-}Z''$ Cardan angles such that the end effector's rotation matrix is formed as $^0\boldsymbol{R}_E(\boldsymbol{x}) = \boldsymbol{R}_x(\varphi_x)\boldsymbol{R}_y(\varphi_y)\boldsymbol{R}_z(\varphi_z)$. The benefit of this selection is, as elaborated in [28], that rotation around the z_E axis – by definition the tool axis – of the end effector corresponds to the last coordinate φ_z. This coordinate does not influence tasks with five DoF and can therefore be eliminated in their task space coordinates $\boldsymbol{y}_E = (r_x, r_y, r_z, \varphi_x, \varphi_y)^{\mathrm{T}}$.

The desired task trajectory is given as a time series of partial poses, i.e. $\boldsymbol{y}_E(t)$, which is sampled discretely as $\boldsymbol{y}_E(t_k)$ with $N + 1$ samples. An at least two times continuous differentiability of the position profile (preferably an S-curve) is required, therefore also $\dot{\boldsymbol{y}}_E(t)$ and $\ddot{\boldsymbol{y}}_E(t)$ exist. Since the trajectory consists of multiple (N_{R}) rest-to-rest phases, at rest times t_{R_i} the condition

$$\dot{\boldsymbol{y}}_E(t_{\mathrm{R}_k}) = 0 \quad \text{for} \quad k = 0, ..., N_{\mathrm{R}} \tag{1}$$

has to hold. Under the assumption of task redundancy, the robot is controlled in it's operational space $\boldsymbol{x}_E$ and only a continuous one-DoF trajectory $\varphi_z(t)$ has to be generated by optimization, since $\boldsymbol{y}_E(t)$ is already given.

2.1 Static Formulation of the Optimization Problem

A static formulation of the optimization problem only for the rest poses at t_{R_k} does not include system dynamics. Therefore the state can be chosen as $x_k = \varphi_z(t_{R_k})$ as the redundant coordinate at rest times. The running cost $l(x_k)$ only depends on the current state. The total cost is then defined as $J = \sum_{k=0}^{N_R} l(x_k)$ and an optimal sequence of states minimizes the total cost by $\boldsymbol{X}^* = \mathrm{argmin}_{\boldsymbol{X}}\, J(\boldsymbol{X}) = \{x_0^*, x_1^*, ..., x_{N_R}^*\}$. This formulation still ignores the transition between the states, which can be included as shown next or by the complete DP formulation of Sect. 4.1.

2.2 Differential Formulation of the Optimization Problem

To incorporate also states between the rest poses, a dynamic formulation can be used, such as differential dynamic programming [33]. The problem can be described by a dynamic system $\boldsymbol{x}_{i+1} = \boldsymbol{f}(\boldsymbol{x}_i, u_i)$, where the state is denoted by $\boldsymbol{x}_i = [\varphi_z(t_i), \dot{\varphi}_z(t_i)]^{\mathrm{T}}$ and the input by $u_i = \ddot{\varphi}_z(t_i)$. The objective is to obtain a trajectory $\{\boldsymbol{X}, \boldsymbol{U}\}$ of all N samples consisting of the control sequences $\boldsymbol{U} = \{u_0, u_1, ..., u_{N-1}\}$ and state sequences $\boldsymbol{X} = \{\boldsymbol{x}_0, \boldsymbol{x}_1, ..., \boldsymbol{x}_N\}$. The final cost is defined as $l_{\mathrm{f}}(\boldsymbol{x}_N)$, the running cost as $l(\boldsymbol{x}_i, u_i)$ and the total cost is

$$J(\boldsymbol{x}_0, \boldsymbol{U}) = \sum_{i=0}^{N} l(\boldsymbol{x}_i, u_i) + l_{\mathrm{f}}(\boldsymbol{x}_N). \qquad (2)$$

The optimal sequence $\boldsymbol{U}^* = \mathrm{argmin}_{\boldsymbol{U}}\, J(\boldsymbol{x}_0, \boldsymbol{U})$ minimizes the total cost. For a rest-to-rest motion additionally the condition

$$\dot{\varphi}_z(t_{R_k}) = 0 \quad \text{for} \quad k = 0, ..., N_R \qquad (3)$$

is regarded to reduce oscillations and allow pausing the complete trajectory motion at rest times. This presents a *phase constraint* between multiple phases according to [13] and complicates using DDP with classical schemes.

2.3 Hypothesis of the Paper: Decoupling of Problem Phases

It is desirable to *decouple* the problem of the single rest-to-rest phases to simplify the overall optimization problem to solve. The decoupling makes it necessary to avoid using a fine discretization of the state x_k at the coupling points (rest times). Otherwise, the coupled problem of Sect. 2.2 arises. The main requirement for the decoupling is the coincidence of global cost of the static problem of Sect. 2.1 and local costs of the rest-to-rest-problem of Sect. 2.2 solved differentially. In other words, the time interdependence of current states and future costs should be low.

By this, locally optimizing the cost also leads to global optimization in general. To avoid local optima, the static optimization of Sect. 2.1 performs an adequate exploration at the rest times.

If in the robot example of this paper e.g. a feasible trajectory is the overall goal (regarding singularity, joint limits and collision avoidance), then this assumption is likely to hold. If the objective is rather a complex relation of the robot state over a long time horizon, the proposed approach will less likely converge to a global optimum. Exemplarily this can be the case for the minimization of the energy consumption over the trajectory, as local increases of kinetic or potential energy may be beneficial depending on later states.

3   Local Optimization: Second-Order Inverse Kinematics

It is assumed that the optimal sequence U_k^* from Sect. 2.2 for rest-to-rest motion k with $t_{R_k} \leq t \leq t_{R_{k+1}}$ can be approximated by local optimization using the locally optimal nullspace projection ("NP") scheme, yielding $U_k^* \approx U_{k,\mathrm{NP}}$. This assumption holds for short trajectories in terms of distance in the performance map. The claim is only qualitative and has to be evaluated for the specific task at hand. In the following, the nullspace projection scheme for the second-order inverse kinematics is wrapped up for serial and parallel robots and extensions are presented which improve using it for global optimization in the next Sect. 4.

3.1   Kinematics Model for Serial and Parallel Robots

Due to their differences in modeling, the kinematics of serial and parallel robots are introduced separately in the following. However, the general relations can be handled interchangeably in the section thereafter on nullspace trajectory motion.

Kinematics of Serial Robots are handled in robotics textbooks like [29]. The position-level forward kinematics present a mapping of joint coordinates q and end effector coordinates $x_E(q)$. The end effector pose[1] x is chosen as presented in Sect. 2. By analytic differentiation the linear velocity and acceleration relation

$$\dot{x} = J_x \dot{q} \quad \text{and} \quad \ddot{x} = \dot{J}_x \dot{q} + J_x \ddot{q} \tag{4}$$

can be obtained, where J_x denotes the analytic manipulator Jacobian, as opposed to the geometric one. The solution to the (non-redundant) inverse differential kinematics is obtained via

$$\ddot{q} = J_x^{-1}(\ddot{x} - \dot{J}_x \dot{q}). \tag{5}$$

---

[1] For the sake of readability, x_E and y_E will be written as x and y in Sect. 3.

Kinematics of Parallel Robots are constructed differently than those for serial robots, since they consists of several kinematic chains, called legs, which connect at a moving platform. The approach to kinematics is the definition of constraints equations $\boldsymbol{\Phi}(\boldsymbol{q}, \boldsymbol{x}) = \boldsymbol{0}$, see e.g. the textbook [17]. Parallel robots' joint coordinates $\boldsymbol{q}$ also contain those of passive joints in the general case, which is regarded in the following. Time differentiation of the constraints leads to

$$\boldsymbol{\Phi}_{\partial q}\dot{\boldsymbol{q}} + \boldsymbol{\Phi}_{\partial x}\dot{\boldsymbol{x}} = \boldsymbol{0} \quad \text{and} \quad \dot{\boldsymbol{q}}_{\mathrm{T}} = -\boldsymbol{\Phi}_{\partial q}^{-1}\boldsymbol{\Phi}_{\partial x}\dot{\boldsymbol{x}} = \boldsymbol{J}_{q,x}^{-1}\dot{\boldsymbol{x}}, \tag{6}$$

where the index "T" denotes the task solution of the differential inverse kinematics (6) as opposed to the nullspace solution "N" discussed later. The index "∂x" denotes the gradient w.r.t. the coordinate $\boldsymbol{x}$ and $\boldsymbol{J}_{q,x}^{-1}$ is the inverse manipulator Jacobian[2] referring to all coordinates. The second time derivative is obtained by differential calculus as

$$\ddot{\boldsymbol{q}}_{\mathrm{T}} = \boldsymbol{J}_{q,x}^{-1}\ddot{\boldsymbol{x}} + \dot{\boldsymbol{J}}_{q,x}^{-1}\dot{\boldsymbol{x}}. \quad \text{with} \quad \dot{\boldsymbol{J}}_{q,x}^{-1} = \boldsymbol{\Phi}_{\partial q}^{-1}(\dot{\boldsymbol{\Phi}}_{\partial q}\boldsymbol{\Phi}_{\partial q}^{-1}\boldsymbol{\Phi}_{\partial x} - \dot{\boldsymbol{\Phi}}_{\partial x}). \tag{7}$$

The manipulator Jacobian $\boldsymbol{J}_x$ can be obtained by selecting the rows in $\boldsymbol{J}_{q,x}^{-1}$ associated to the active joint's coordinates $\boldsymbol{\theta} = \boldsymbol{P}_\theta\boldsymbol{q}$ and matrix inversion by

$$\boldsymbol{J}_x = \left(\boldsymbol{J}_x^{-1}\right)^{-1} = \left(\boldsymbol{P}_\theta\boldsymbol{J}_{q,x}^{-1}\right)^{-1}. \tag{8}$$

The forward differential equations from actuator to end effector velocities and accelerations can then be expressed like in (4) as

$$\dot{\boldsymbol{x}} = \boldsymbol{J}_x\dot{\boldsymbol{\theta}} \quad \text{and} \quad \ddot{\boldsymbol{x}} = \dot{\boldsymbol{J}}_x\dot{\boldsymbol{\theta}} + \boldsymbol{J}_x\ddot{\boldsymbol{\theta}}. \tag{9}$$

If actuator entities are given, they can be transferred to the full joint space by

$$\dot{\boldsymbol{q}} = \boldsymbol{J}_{q,x}^{-1}\boldsymbol{J}_x\dot{\boldsymbol{\theta}} \quad \text{and} \quad \ddot{\boldsymbol{q}} = \boldsymbol{J}_{q,x}^{-1}\boldsymbol{J}_x\ddot{\boldsymbol{\theta}} + \boldsymbol{J}_{q,x}^{-1}\dot{\boldsymbol{J}}_x\dot{\boldsymbol{\theta}} + \dot{\boldsymbol{J}}_{q,x}^{-1}\boldsymbol{J}_x\dot{\boldsymbol{\theta}}. \tag{10}$$

3.2   Second-Order Nullspace Inverse Kinematics Controller Scheme

In the following, all relations hold for serial and parallel robots simultaneously. Since all joints of serial robots are assumed as active, $\boldsymbol{\theta} := \boldsymbol{q}$ holds in that case. The inversion of the kinematics equations provides a nullspace solution if a task redundancy exists. The transfer from operational space $\boldsymbol{x}$ to task space $\boldsymbol{y}$ is simplified by the chosen X-Y'-Z''-angle representation such that the last row of (4) and (9) can be removed with $\boldsymbol{y} = \boldsymbol{P}_y\boldsymbol{x}$ and $\boldsymbol{J}_y = \boldsymbol{P}_y\boldsymbol{J}_x$ (for serial and for parallel robots). If no specific value for the redundant coordinate is given, the minimum-norm solution for the task motion can be obtained using the pseudo inverse † as

$$\ddot{\boldsymbol{\theta}}_{\mathrm{T,minnorm}} = \boldsymbol{J}_y^\dagger(\ddot{\boldsymbol{y}} - \dot{\boldsymbol{J}}_y\dot{\boldsymbol{\theta}}). \tag{11}$$

---

[2] The expression is not a regular inverse, since it is not square and the non-inverse does not exist. The notation is used to unify the symbols with those of serial robots.

The nullspace solution of the inverse kinematics on acceleration level is obtained as the homogenous solution of (11), neglecting the term $\dot{\boldsymbol{N}}_\theta$, as

$$\ddot{\boldsymbol{\theta}}_N = (\boldsymbol{I} - \boldsymbol{J}_y^\dagger \boldsymbol{J}_y)\boldsymbol{v}_\theta = \boldsymbol{N}_\theta \boldsymbol{v}_\theta, \tag{12}$$

which projects arbitrary vectors $\boldsymbol{v}_\theta$ into the nullspace. The full expression for the actuator accelerations can thus be combined from $\ddot{\boldsymbol{\theta}}_T$ from (5)/(7) (serial/parallel) or (11) in combination with $\ddot{\boldsymbol{\theta}}_N$ from (12) as

$$\ddot{\boldsymbol{\theta}} = \begin{cases} \ddot{\boldsymbol{\theta}}_T + \ddot{\boldsymbol{\theta}}_N & \text{for a given } \varphi_z = \varphi_{z,\text{ff}} \\ \ddot{\boldsymbol{\theta}}_{T,\text{minnorm}} + \ddot{\boldsymbol{\theta}}_N & \text{otherwise.} \end{cases} \tag{13}$$

In the following, the vector $\boldsymbol{v}_\theta$ for the second-order nullspace projection (12) is created by a PD controller scheme [3,28] such that a stable nullspace optimization is performed. This substitutes the analytic scheme of [24] numerically. By using $h_{\partial\theta}$ as a feedback for $\boldsymbol{v}_\theta$ the performance criterion h is optimized, as discussed in the next subsection. The overall nullspace controller is shown in the right part of Fig. 2 and contains the controller with PD gains K_P, K_D and additional damping K_v. The trajectory input from (13) is placed on the left part of the figure and allows switching between the cases whether a feedforward term $\ddot{\varphi}_{z,\text{ff}}$ for the redundant coordinate is given or not. In the first case, the nullspace motion develops around the feedforward trajectory $\varphi_{z,\text{ff}}(t)$ and in case of $h = \text{const}$ the redundant coordinate will move according to it.

The model requires computing matrices $\boldsymbol{J}_{q,x}^{-1}$ and $\dot{\boldsymbol{J}}_{q,x}$ which depend on the full robot state $\boldsymbol{q}, \dot{\boldsymbol{q}}, \boldsymbol{x}, \dot{\boldsymbol{x}}$, shown by the upper feedback branch. The actual value φ_z of the redundant coordinate has to be computed via the position-level inverse kinematics and the velocity $\dot{\varphi}_z$ using the manipulator Jacobian with (4) or (9).

3.3   Performance Criteria for Gradient Projection

For a feasible motion strict constraints regarding self collision, joint limits and singularities have to be met, which are implemented as objective within the nullspace optimization. Optimizing multiple criteria h_i with only one degree of freedom does not allow using task priority schemes like [14,21]. The drawback of a weighted sum $h = \sum w_i h_i$ as nullspace objective is that prioritization is performed by the weights w_i, which lack of a physical meaning and often have to be

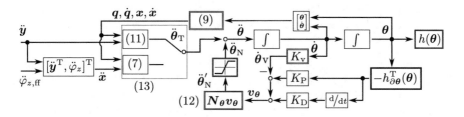

Fig. 2. Nullspace motion controller scheme with nullspace trajectory feedforward.

tuned manually. To encounter the weights-tuning problem, optimization criteria to achieve feasibility constraints were chosen that lead to infinite penalty, following e.g. [35]. The multi-objective problem is encountered by choosing objective functions with an activation threshold to avoid permanent activity of multiple criteria. This can be seen as a special case of set-based task-priority frameworks like [14].

Joint Limits are regarded with the hyperbolic joint limit criterion

$$h_{\text{lim}}(\boldsymbol{q}) = \frac{1}{\dim(\boldsymbol{q})} \sum_{i=1}^{\dim(\boldsymbol{q})} h_{\text{lim}}(q_i) \quad \text{with} \quad h_{\text{lim}}(q_i) = h_{\text{lim,hyp}}(q_i) \quad \text{and} \quad (14)$$

$$h_{\text{lim,hyp}}(q_i) = \frac{(q_{i,\max} - q_{i,\min})^2}{8} \left(\frac{1}{(q_i - q_{i,\min})^2} + \frac{1}{(q_i - q_{i,\max})^2} \right) \geq 1 \quad (15)$$

from [35] with the modification of an activation threshold. The criterion is only active if the limit is approached passing a lower or upper threshold $q_{i,\text{thr,min}}$ or $q_{i,\text{thr,max}}$. The continuous differentiability is achieved via cubic spline transitions and switching points $q_{i,\text{sw,min}}$ and $q_{i,\text{sw,max}}$, resulting in the criterion

$$h_{\text{lim}}(q_i) = \begin{cases} \infty & \text{for} \quad q_i < q_{i,\min} \quad \text{(lower limit)} \\ h_{\text{lim,hyp}}(q_i) & \text{for} \quad q_{i,\min} \leq q_i < q_{i,\text{sw,min}} \\ h_{\text{spline,ll}}(q_i) & \text{for} \quad q_{i,\text{sw,min}} \leq q_i < q_{i,\text{thr,min}} \\ 0 & \text{for} \quad q_{i,\text{thr,min}} < q_i < q_{i,\text{thr,max}} \quad \text{(inactive)} \\ h_{\text{spline,ul}}(q_i) & \text{for} \quad q_{i,\text{thr,max}} \leq q_i < q_{i,\text{sw,max}} \\ h_{\text{lim,hyp}}(q_i) & \text{for} \quad q_{i,\text{sw,max}} \leq q_i \leq q_{i,\max} \\ \infty & \text{for} \quad q_{i,\max} < q_i \quad \text{(upper limit).} \end{cases} \quad (16)$$

The gradient $h_{\partial q} = \partial h / \partial \boldsymbol{q}$ can be obtained analytically. In the paper's examples, $q_{i,\text{thr}}$ is set to be 90% of the limit range and $q_{i,\text{sw,max}} = (q_{i,\text{thr,max}} + q_{i,\max})/2$.

Platform Coordinate Limits have to be restricted for the global optimization algorithm presented in Sect. 4. The redundant coordinate's range is $\varphi_{z,\min} \leq \varphi_z \leq \varphi_{z,\max}$, also ensured by nullspace optimization and a criterion $h_{\varphi_z,\text{lim}}$ analogue to (16) with φ_z instead of q_i. The limit $\varphi_{z,\min}$ can be adjusted over time around a given feedforward reference $\varphi_{z,\text{ff}}(t)$ to enlarge or diminish the allowed range of the redundant coordinate, as shown in Sect. 4 by spline interpolation.

Singularities are avoided by means of nullspace optimization, where the condition number $h_{\text{cond,II}} = \text{cond}(\boldsymbol{J}_x)$ of the manipulator Jacobian is used directly as objective. Inconsistent units of the condition number and questionable physical meaning [12,17] are ignored since an infinite numerical value represents a singularity in any case. To allow optimization of other criteria as well, an activation threshold $h_{\text{cond,act}}$ for the singularity criterion is defined and a cubic spline transition h_{spline} leads to direct use of the condition number after the threshold $h_{\text{cond,thr}}$, resulting in

$$
h_{\text{sing}} = \begin{cases} 0 & \text{for} \quad h_{\text{cond}} < h_{\text{cond,act}} \quad \text{(inactive)} \\ h_{\text{spline}}(h_{\text{cond}}) & \text{for} \quad h_{\text{cond,act}} \leq h_{\text{cond}} < h_{\text{cond,thr}} \\ h_{\text{cond}} & \text{otherwise} \quad \text{(reaches } \infty \text{ in singularity).} \end{cases} \tag{17}
$$

For obtaining the gradient $h_{\partial q} = \partial h/\partial q$, a numeric implementation via difference quotients is used, as detailed in [28]. In the further examples of this paper, $h_{\text{cond,II,act}} = h_{\text{cond,II,thr}} = 1$ leads to a permanent optimization of the condition number, which has a weak (but questionable) correlation with other performance criteria [17].

Additionally, singularities of the inverse kinematics of $\boldsymbol{\Phi}_{\partial q}$ for parallel robots and $\boldsymbol{J}_y$ for serial robots are handled in the same manner. This allows to avoid serial (type I, $\boldsymbol{\Phi}_{\partial q}$) and parallel (type II, $\boldsymbol{J}_x$) singularities for parallel robots. The respective criteria are termed $h_{\text{sing,I}}$ and $h_{\text{sing,II}}$.

Self-collisions of the robot structure are considered by a simplified convex geometric model of the robot. The six links are represented by capsules (cylinders ending in half-spheres) to allow a fast geometric check for body intersection. The platform is modeled as a ring of six capsules, resulting in 39 elementary collision checks with a minimal distance $d_{\text{coll,min}} = \min d_i$ of any of the objects. An intersection leads to a negative value. The collision criterion

$$
h_{\text{coll}}(d_{\text{coll,min}}) = \begin{cases} 0 & \text{for} \quad d_{\text{coll,min}} \geq d_{\text{coll,thr}} \quad \text{(safe distance)} \\ h_{\text{spline}}(d_{\text{coll,min}}) & \text{for} \quad d_{\text{thr,hyp}} \leq d_{\text{coll,min}} < d_{\text{coll,thr}} \\ h_{\text{coll,hyp}}(d_{\text{coll,min}}) & \text{for} \quad 0 < d_{\text{coll,min}} < d_{\text{thr,hyp}} \\ \infty & \text{for} \quad d_{\text{coll,min}} \leq 0 \quad \text{(collision)} \end{cases} \tag{18}
$$

is activated if a safety threshold $d_{\text{coll,thr}}$ is exceeded. A cubic spline sets the transition from zero to one branch of a hyperbola similar to (15), beginning at distance $d_{\text{thr,hyp}}$, which reaches ∞ in case of collision. The gradient is computed numerically via difference quotients.

The Combined Performance Criterion is obtained by a weighted sum

$$
h = w_{\text{sing,I}}h_{\text{sing,I}} + w_{\text{sing,II}}h_{\text{sing,II}} + w_{\text{lim}}h_{\text{lim}} + w_{\text{coll}}h_{\text{coll}} + w_{\varphi_z,\text{lim}}h_{\varphi_z,\text{lim}}, \tag{19}
$$

where $w_{\text{sing,I/II}} = 1$, $w_{\text{lim}} = 1$, $w_{\text{coll}} = 1$ and $w_{\varphi_z,\text{lim}} = 100$ are chosen by manual tuning to obtain a sufficiently strong repulsion from the coordinate limits for the example in Sect. 4. The gains $K_{\text{P}} = 1$, $K_{\text{D}} = 0.7$ and $K_{\text{v}} = 0.8$ were chosen to reduce oscillations of the second-order system resulting from the PD feedback of a double integrator plant. For visualization of h within a performance map, high values above a threshold of e.g. $h_{\mathrm{thr}} = 1000$ are saturated to increase the number of heatmap colors for distinction of good values. The component of h leading to infinity can then be highlighted by a separate marker.

3.4   Extensions for Rest-to-Rest Motion

To achieve a rest-to-rest motion of the robot trajectory $\boldsymbol{x}(t)$, an additional braking has to be implemented within the nullspace motion $\varphi_z(t)$. Otherwise, a nullspace velocity $\dot{\varphi}_z$ is likely to still persist at the task's resting points. As elaborated before in Sect. 2 and (1), rest-to-rest motion for the task trajectory $\boldsymbol{y}(t)$ is assumed as given. A time-varying technically feasible limit $\dot{\varphi}_{z,\mathrm{min}}(t)$ and $\dot{\varphi}_{z,\mathrm{max}}(t)$ for the redundant coordinate's velocity is set. When the resting time t_{R} is approached with $t > t_{\mathrm{R}} - T_{\mathrm{dec}}$, the limit is linearly decreasing to zero in the deceleration time T_{dec}, which is obtained from acceleration and velocity limits. The nullspace braking is implemented within the saturation block in Fig. 2. A predicted velocity from the task and nullspace acceleration for the next discrete time step is computed by $\dot{\boldsymbol{\theta}}_{\mathrm{pre}}(k+1) = \dot{\boldsymbol{\theta}}(k) + (\ddot{\boldsymbol{\theta}}_{\mathrm{T}} + \ddot{\boldsymbol{\theta}}_{\mathrm{N}}')\Delta t$. The redundant coordinate's velocity is obtained by the Jacobian relation $\dot{\varphi}_{z,\mathrm{pre}} = \boldsymbol{P}_{\varphi_z}\boldsymbol{J}_x\dot{\boldsymbol{\theta}}_{\mathrm{pre}}(k+1)$, where $\boldsymbol{P}_{\varphi_z} = [\boldsymbol{0}^{\mathrm{T}}, 1]$ selects the last row. If $\dot{\varphi}_{z,\mathrm{pre}}$ exceeds the limit $\dot{\varphi}_{z,\mathrm{max}}$, an additional nullspace acceleration $\ddot{\varphi}_{z,\mathrm{add}} = (\dot{\varphi}_{z,\mathrm{max}} - \dot{\varphi}_{z,\mathrm{pre}})/\Delta t$ is created to truncate the velocity. This acceleration is then added to the robot's joint space by $\ddot{\boldsymbol{\theta}}_{\mathrm{N}} = \ddot{\boldsymbol{\theta}}_{\mathrm{N}}' + \boldsymbol{J}_x^{-1}[\boldsymbol{0}^{\mathrm{T}}, \ddot{\varphi}_{z,\mathrm{add}}]^{\mathrm{T}}$.

Additionally, a damping is added in the nullspace, which promotes following the redundant coordinate's feedforward $\dot{\varphi}_{z,\mathrm{ff}}$. The block K_{v} in Fig. 2 is implemented by a feedback law $\dot{\boldsymbol{\theta}}_{\mathrm{v}} = K_{\mathrm{v}}\boldsymbol{P}_{\varphi_z}\boldsymbol{J}_x(\dot{\varphi}_z - \dot{\varphi}_{z,\mathrm{ff}})$ of the redundant coordinate's velocity.

3.5   Trajectory Performance Criterion

To obtain a criterion for evaluation of the complete rest-to-rest trajectory k, an RMS criterion w.r.t. the normalized path coordinate s is defined as

$$h_{\mathrm{int}} = \frac{1}{s_{\mathrm{R}_k} - s_{\mathrm{R}_{k-1}}} \int_{s_{\mathrm{R}_{k-1}}}^{s_{\mathrm{R}_k}} h(\boldsymbol{q}(s), \boldsymbol{x}_E(s))\mathrm{d}s. \tag{20}$$

The path coordinate s is used instead of time t to reduce the weight of sections with low velocity. It has integer values at the rest times t_{R_k} and interpolates linearly according to the distance of the task coordinate $\boldsymbol{y}$. The use of the root mean square (RMS) in (20) and the definition of h in (19) make it susceptible to the nearly infinite penalty values when a singularity or a limit is approached. This desirable behavior allows the simultaneous incorporation of the constraints and the objective in the optimization of rest poses via dynamic programming, which is the focus of the next section.

4 Global Optimization: Dynamic Programming

The local optimization using nullspace projection from Sect. 3 is now used for the global trajectory optimization problem introduced in Sect. 2. First, the problem is solved with the well-known (discrete) dynamic programming method in Sect. 4.1 which is then extended to a novel state-interval-based dynamic programming method in Sect. 4.2.

4.1 Discrete Dynamic Programming

The stage-wise definition of the rest-to-rest trajectory optimization problem is

$$J^* = \min_U J(u_1, ..., u_n) = \sum_{k=1}^{n} l_k(x_{k-1}, u_k) \tag{21}$$

$$\text{s.t.} \quad x_k = f_k(x_{k-1}, u_k) \tag{22}$$

$$x_k \in X_k \tag{23}$$

$$u_k \in U_k(x_{k-1}). \tag{24}$$

The total cost J is a sum of costs l_k on each of the n stages, which are only influenced by decisions u_k on that stage and the stage's initial state x_{k-1}. The initial value x_0 is given. The goal is to find a trajectory of decisions $U = \{u_0, u_1, u_{n-1}\}$ as argmin in (21). Optimal values are marked with an asterisk, like J^*. Only a discrete set of states X_k is possible (23). The transfer between states is denoted as function f_k in (22) which produces the transfer costs l_k. Decisions are restricted to a set U_k leading to these exact states (24). The set of states is predetermined as a discretization of the continuous decision variable x for n_{ref} reference values as

$$X_{\text{ref}} = \{x_{\text{ref},1}, x_{\text{ref},2}, ..., x_{\text{ref},n_{\text{ref}}}\}. \tag{25}$$

In a forward iteration, the running costs are obtained for in total $|X_{k-1}||U_k|$ different transfers from stage $k-1$ to stage k, termed by

$$l_k(x_{k-1}, u_k) \quad \forall \quad x_{k-1} \in X_{k-1} \wedge u_k \in U_k(x_{k-1}). \tag{26}$$

The cumulated cost J_k for each of the states x_k of stage k is then obtained via

$$J_k(x_{k-1}, u_k) = J_{k-1}^*(x_{k-1}) + l_k(x_{k-1}, u_k), \tag{27}$$

where $J_{k-1}^*(x_{k-1})$ is the optimal total cost from x_0 to x_{k-1}, initialized with $J_0^*(x_0) = 0$. The optimal series of decisions for each state x_k on stage k is obtained via considerations based on Bellman's principle of optimality as

$$J_k^*(x_k) = \min\{J_k(x_{k-1}, u_k) \mid x_k = f(x_{k-1}, u_k) \wedge u_k \in U_k(x_{k-1})\}. \tag{28}$$

All predecessor states x_{k-1} are considered. The set $U_k(x_{k-1})$ of decision variables is selected such that each of the reference states $X_k := X_{\text{ref}}$ is reached once. If the state x_{k-1} can only be reached by violation of a constraint, this is marked by the previous iteration with infinite cost $J_k^*(x_{k-1})$ for that state, resulting to

$$
U_k(x_{k-1}) = \begin{cases} \emptyset & \text{for } J_k^*(x_{k-1}) = \infty \\ \{u_k \mid x_k = f(x_{k-1}, u_k) \ \forall \ x_k \in X_k\} & \text{otherwise.} \end{cases} \quad (29)
$$

The optimal series of decisions and states leading to $J_k^*(x_k)$ is written as

$$
\begin{aligned}
U^*(x_k) &= \{u_1^*, ..., u_k^*\} \quad \text{and} \\
X^*(x_k) &= \{x_0^*, x_1^*, ..., x_{k-1}^*, x_k^*\}.
\end{aligned} \quad (30)
$$

The dynamic programming algorithm now consists of alternating forward passes (26) and backward passes (28) to select optimal stage decisions. The algorithm is performed for all stages from $k = 1, ..., n$. The best final state then contains the optimal solution as

$$
J^* = J_k^*(x_n^*) = \min J_k^*(x_n) \quad (31)
$$
$$
U^* = U^*(x_n^*) = \{u_1^*, ..., u_n^*\} \quad (32)
$$
$$
X^* = X^*(x_n^*) = \{x_0^*, x_1^*, ..., x_n^*\} \quad (33)
$$

Robot Trajectory Example. The general form of the algorithm is transferred to the robot example in the following by assigning the DP variables from above with the physical variables in Sect. 2 and 3. The stages correspond to the rest positions of the trajectory. In the example of [28] depicted in Fig. 1 this means $n = N_R = 7$. Each transition $x_k = f(x_{k-1}, u_k)$ is a rest-to-rest trajectory $\varphi_z(t)$ for the redundant coordinate computed by a trapezoidal velocity profile for the given trajectory time base $t_{R_{k-1}} \leq t \leq t_{R_k}$ between given states in $x_{k-1} = \varphi_z(t_{R_{k-1}})$ and $x_k = \varphi_z(t_{R_k})$. The decision u_k is therefore just the selection of a target position x_k. The discretization is chosen as $x_{min} = \varphi_{z,min} = -180°$, $x_{max} = \varphi_{z,max} = 180°$ and $n_{ref} = 9$, leading to $\Delta x = \Delta \varphi_z = 45°$. The cost is determined by the RMS of the performance criterion (20), according to Sect. 3.3, i.e.

$$
l_k(x_{k-1}, u_k) = \frac{1}{s_{R_k} - s_{R_{k-1}}} \int_{s_{R_{k-1}}}^{s_{R_k}} h(q(s), x_E(s)) dt. \quad (34)
$$

The initial value $x_0 = \varphi_{z,0} \approx -35°$ and $q(t = 0)$ is selected by a gradient-descent approach for the optimal joint limit criterion. The joint configuration $q(t)$ is obtained by the inverse kinematic scheme from Sect. 3.

A step-by-step solution with the dynamic programming approach above is shown in Fig. 3a for the first stage $k = 1$. As only one previous stage $X_0 = \{x_0\}$ exists, the only one transfer to each x_k from (30) is the optimal transfer. The trajectory evaluation is immediately aborted if a robot constraint is violated. This speeds up the algorithm and is marked by black lines in Fig. 3. The corresponding cost l_k is set to a high penalty (in this case ∞) to discard this state automatically in the backward recursion (28). If all decisions towards a state x_{k-1} are invalid, the first condition in (29) becomes active and from this state

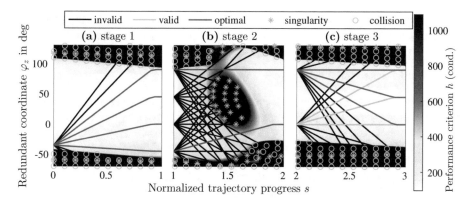

Fig. 3. Decisions on the first three stages using discrete DP for the reference problem.

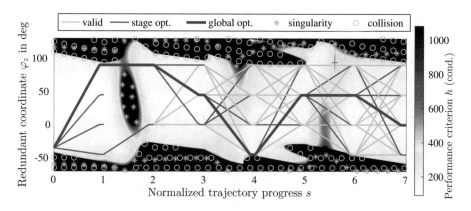

Fig. 4. Result of discrete dynamic programming for the reference problem.

no actions will be taken on further. The forward iteration for the next stage $k = 2$ therefore only continues for valid states. Only one trajectory for the first state and the last state give feasible transfers, as visible in Fig. 3b. The recursion therefore remains trivial. In iteration $k = 3$ both remaining states from X_2 lead to feasible transfers to the three states in X_3, as can be seen in Fig. 3c. Further, each state in X_3 can be reached by both states of X_2, from which the upper one has lower cost and is therefore marked as optimal (magenta). Lines originating from the lower state are marked as valid (cyan), but not optimal.

The final result of the algorithm after performing the remaining stages $k = 4...7$ is shown in Fig. 4 with the highlighted optimal trajectory that has been found. The continuous optimization variable can only very roughly be approximated by the chosen discretization of $\varphi_z = 45°$. Choosing a finer discretization is able to improve the result in this case, but using the discrete DP is not applicable to the general case, if a feasible solution can not be found via straight lines in the performance map. Further, the DP algorithm has complexity $\mathcal{O}(nn_{\text{ref}}^2)$, making the presented implementation inefficient for such continuous problems.

4.2   State-Interval Dynamic Programming

The drawbacks of the classical discrete dynamic programming can be solved by defining a target interval for the states rather than a fixed value. A state interval

$$[x] = [\bar{x} - {}^{\Delta x}\!/_2, \bar{x} + {}^{\Delta x}\!/_2] \tag{35}$$

is defined by a center $\bar{x}$ and the interval width Δx. Instead of a fixed set of discretized reference states X_{ref} in (25), the states are now[3]

$$X_{\text{ref}} = \{[x_{\text{ref},1}], [x_{\text{ref},2}], ..., [x_{\text{ref},n_{\text{ref}}}]\}. \tag{36}$$

The DP formulation from the previous section is adapted in such a way that previous states remain specific values x_{k-1}. Future states are considered as an interval, allowing a set of solutions $x_k \in [x_k]$ or $\bar{x}_k - {}^{\Delta x}\!/_2 \leq x_k < \bar{x}_k + {}^{\Delta x}\!/_2$. The set of decision variables from (29) now becomes

$$U_k(x_{k-1}) = \begin{cases} \emptyset & \text{for } J_k^*(x_{k-1}) = \infty \\ \{u_k \mid f(x_{k-1}, u_k) \in [x_k] \ \forall \ [x_k] \in X_k\} & \text{otherwise.} \end{cases} \tag{37}$$

Since the intervals do not overlap, Bellman's principle of optimality still holds under the assumption of an optimal state transfer regarding the subproblem in $f(x_{k-1}, u_k)$. This means that the transfer to the actual reached state x_k within the interval $[x_k]$ is assumed to be optimal among all possible local transfer strategies $\tilde{u}_k$, i.e.

$$l_k(x_{k-1}, u_k) = \min_{\tilde{u}_k} \{l_k(x_{k-1}, \tilde{u}_k) \ \forall \ \tilde{u}_k \mid f(x_{k-1}, \tilde{u}_k) \in [x_k]\}. \tag{38}$$

The transition strategy u_k in this case is not only correspondence to a target state x_k, but also a specific control strategy for the subproblem $f(x_{k-1}, u_k)$. In the trajectory optimization example this includes the optimal decision for each of the dense trajectory samples, as mentioned in Sect. 2.2. The assumption (38) is unlikely to strictly hold in the presented highly nonlinear robot application. However, approximating this by a near-optimal or at least feasible u_k may already be enough to obtain acceptable results regarding difficulties in finding valid solutions at all, facing the constraints.

The decision u_k now only has to assure that the next state lies in it's allowed interval. This improves the possibility to perform a local optimization in the state transfer function, which corresponds to the trajectory optimization via the

---

[3] Intervals are assumed non-overlapping (i.e. half-open) for the sake of mathematical proof, however the symbol $[\cdot]$ for closed intervals is used to enhance readability.

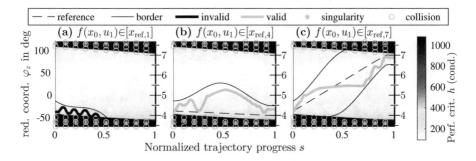

Fig. 5. Decisions u_1 on the first stage using state-interval dynamic programming to exemplary state intervals 1 **(a)**, 4 **(b)** and 7 **(c)** on stage 2. The numbers i on the axis to the right of each performance map correspond to the state intervals $[x_{\text{ref},i}]$.

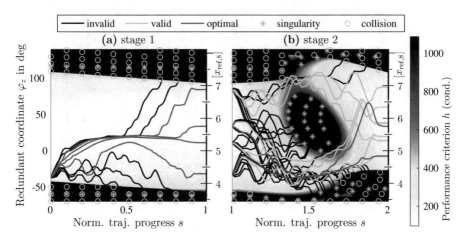

Fig. 6. All decisions on the first two stages using state-interval DP.

nullspace projection. Reaching the interval has to be implemented with extensions to the local optimization or with constraints, as discussed in Sect. 3.3.

The backward recursion from (28) using the interval approach now is

$$J_k^*([x_k]) = \min\left\{J_k(x_{k-1}, u_k) \mid f(x_{k-1}, u_k) \in [x_k] \wedge u_k \in U_k(x_{k-1})\right\}. \quad (39)$$

The cumulated cost $J_k(x_{k-1}, u_k)$ does not change compared to (27), since looking retrospectively, the previous state x_{k-1} is the actual obtained value, no interval.

The proposed approach is illustrated at the previous robot trajectory example. Interval limits are set as before as $x_{\text{ref},1} = \varphi_{z,\min} = -180°$, $x_{\text{ref},n_{\text{ref}}} = \varphi_{z,\max} = 180°$, $n_{\text{ref}} = 9$, $\Delta x = \Delta\varphi_z = 45°$ and $x_0 = \varphi_{z,0} \approx -35°$. The term $x_{\min}$ and $x_{\max}$ is not used since the limits for the optimization variable are extended by the interval half-span $\Delta x/2$. A decision u_k consists of a reference trajectory as in the previous discrete DP example. This trajectory is only tracked via a feed-forward gain and a nullspace controller, as explained in Sect. 3. A tolerance

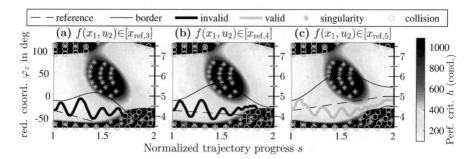

Fig. 7. Decisions u_2 on the second stage from $x_1 \in [x_{\text{ref},4}]$ using state-interval DP.

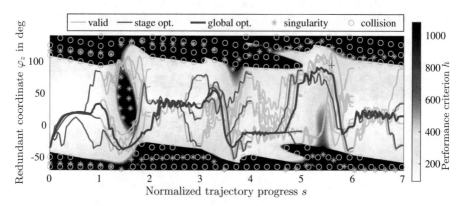

Fig. 8. Result of state-interval dynamic programming for the reference problem.

band around the trajectory ensures reaching the respective target interval $[x_k]$, as shown in Fig. 5 for three exemplary decisions on the first stage.

As in the previous example, not all target states lead to valid results, especially since the global distribution of the performance map is not known in advance and therefore a wide span is chosen. All transitions of the first stage are shown in Fig. 6a. The valid lines are continued in the next stage. Since only one starting state exists, all valid lines are also optimal, which is highlighted by the color magenta. The continuation for stage 2 is shown first exemplarily together with the reference trajectory and spline-based tolerance intervals in Fig. 7. Figure 6b contains all transfers. Multiple decisions u_1 lead to the same state intervals of $[x_2]$, as visible for the cyan lines ending in $[x_{\text{ref},5}]$ and $[x_{\text{ref},6}]$ (noted by the numbers at the right side of Fig. 6b). Therefore the best decisions according to (39) are selected for continuation, marked magenta. The result of the algorithm is depicted in Fig. 8.

4.3 Overlapping Intervals

The approach presented above enforces the redundant coordinate in a tolerance band by using a repulsing potential $h_{\varphi_z,\text{lim}}$. This has the effect that a local optimum on this border between intervals can not be reached. A solution to achieve this is the use of overlapping intervals. It still has to be assured, that at each stage only the prescribed number of state intervals are continued. Otherwise the number of states will grow from state to state and the underlying optimality principle of dynamic programming does not hold any more. The extension leads to additional possible decisions on each set, extending the set X_k in (37) to

$$X_k' = \{[x_{\text{ref},1}], [x_{\text{ref},2}], ..., [x_{\text{ref},n_{\text{ref}}}], [x_{\text{add},1}], ..., [x_{\text{add},n_{\text{add}}}]\}. \tag{40}$$

The additional intervals are set to be overlapping with the existing ones, i.e.

$$[x_{\text{add},k}] = [x_{\text{ref},k}] + \Delta x/2 = x_{\text{ref},k} \leq x < x_{\text{ref},k+1} \tag{41}$$

and $n_{\text{add}} = n_{\text{ref}} - 1$. To prevent the number of states from increasing, the cumulated cost in the backward recursion (39) is not evaluated for the new, overlapping intervals from (40).

Robot Trajectory Example. The modification of overlapping intervals is again demonstrated at the robot example. The chosen parameters are now $n_{\text{ref}} = 5$ with $\Delta x = \Delta\varphi_z = 90°$. Due to the overlapping intervals, the number of forward iterations stays the same, but fewer states are considered for continuation. Similar to Fig. 5, the first stage transfer is investigated in Fig. 10. The transfer in Fig. 10c corresponds to the additional interval $[x_{\text{add},3}]$ between $[x_{\text{ref},3}]$ in Fig. 10a and $[x_{\text{ref},4}]$ in Fig. 10b.

Similar to Fig. 6 all transfers on the first two stages are shown in Fig. 9. Transfers from additional overlapping intervals are marked with dashed lines. Now already in the first stage in Fig. 9a multiple actions lead to a transfer to the interval $[x_{\text{ref},3}]$, which seemingly contains the local optimum. The best of these is selected in stage 2 (Fig. 9b) for continuation.

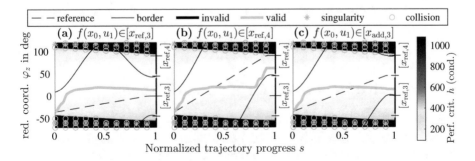

Fig. 9. Decisions u_1 on the first stage using overlapping state-intervals in dynamic programming to exemplary state intervals 3 (**a**), 4 (**b**) and the overlapping interval in-between (**c**) on stage 2.

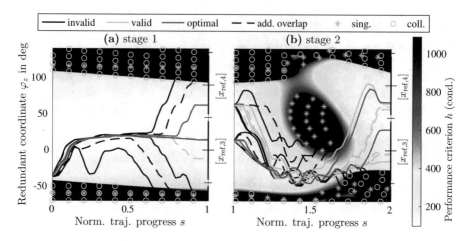

Fig. 10. Decisions on stage 1 and 2 for the reference problem using overlapping intervals.

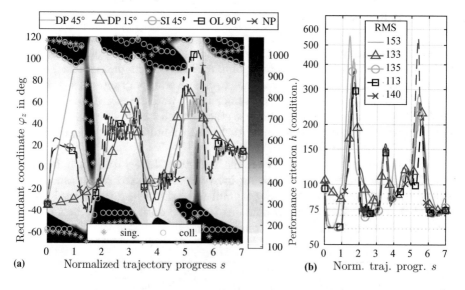

Fig. 11. Nullspace motion in performance map **(a)** and evolution of criterion **(b)**.

5   Simulative Validation for a Parallel Robot

In the following a quantitative evaluation of the proposed new method of state-interval dynamic programming (SI-DP) from Sect. 4.2 is performed with a comparison against classical dynamic programming (DP) from Sect. 4.1 and the previous approach of local optimization using nullspace projection (NP) from Sect. 3 (and [28]). An overview of robot and task can be obtained from Fig. 1.

The same benchmark task as in the previous section is used. The robot has the following dimensions: base diameter 1200 mm, platform diameter 300 mm and distance of hexagonally aligned platform coupling joint pairs of 100 mm. Collision bodies for the leg chains have a diameter of 40 mm. The task is beginning at $[r_x, r_y, r_z] = [-50, 40, 700]$ mm and $[\varphi_x, \varphi_y] = [45°, 0°]$ in the robot base frame and has a length of 900 mm and duration of 31.6 s with seven rest poses. The trajectory has 31647 samples with a sample time of 1 ms.

The comparison in Fig. 11 shows that the SI-DP with a discretization of 45° outperforms the classical discrete DP with the same discretization in the critical phases of the trajectory. Further, DP presents a different trajectory with $\varphi_z = 90°$ at $s = 2$ due to the narrow passage. The reason can be found in the only very rough discretization, which is beneficial for the SI-DP and reduces computation time. A further improvement is achieved by using the overlapping (OL) intervals from Sect. 4.3 with discretization of 90°. The evaluation of one trajectory sample needs 0.4 ms for the DP and 1.2 ms for the SI-DP and OL-SI-DP, due to more complex nullspace equations. A Linux desktop computer with Intel i5-7500 CPU and MATLAB implementation using MEX-compiled functions was used. The total optimization for DP with 45° took 3.4 min with 535k trajectory samples, corresponding to 16.9 times the full trajectory length. The SI-DP with 45° took 13.2 min for 664k samples, i.e. 21 full trajectory equivalents and the OL-SI-DP took 10 min for 421k samples. The RMS value according to (20) is given in Fig. 11b and shows the improved performance by the SI-DP by a value of 135 and OL by 113. The DP results can be improved by finer discretization of 15°, which leads to a similar RDP value of 134 like SI-DP, instead of 153, but takes 25.7 min for 4M samples, i.e. 128 full trajectory equivalents.

The NP method performs better in the first two phases since the optimal solution is between intervals of the SI-DP, which is a repulsing potential. Using the OL method solves this problem. The only local optimality of NP becomes visible at $s = 5$ where a local optimum is followed by a region of high cost terms, which has to be traversed. This leads to a moderately worse RMS value of 140, but needs the least computational effort with 39 s for one full trajectory.

The example is modified in the following to pose higher restrictions on the robot. The trajectory length is increased to 1400 mm and the pointing direction now is 45° to the outside, i.e. $[r_x, r_y, r_z] = [-200, 150, 700]$ mm and $[\varphi_x, \varphi_y] = [-45°, 0°]$ for the starting point. The robot is shown in Fig. 12 for three different points of this second trajectory. The cases of singularity and collision are highlighted for illustration. The trajectory optimization is performed with the same methods and settings as before with results presented in Fig. 13. Due to the very narrow passages of possible rotation angle φ_z, the 45°-DP does not find a solution, as well as the local optimization. The fine-discretization 15°-DP and the proposed 45°-SI-DP find a solution, where SI-DP outperforms the DP by means of RMS of the condition number and OL performs best due to less constraints for enforcing the interval.

For both examples oscillations can be observed in Fig. 11 and Fig. 13 for the nullspace optimization methods SI-DP, OL-SI-DP and NP. This could be

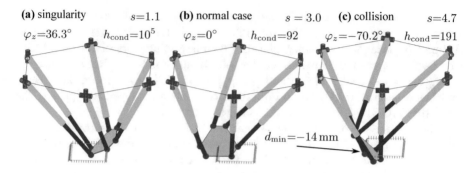

Fig. 12. Robot in three poses of the second performance map example from Fig. 13.

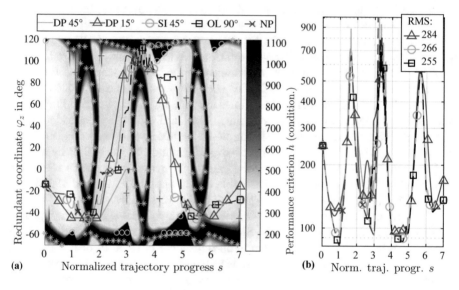

Fig. 13. Results for the second example: Performance map (**a**) and criterion (**b**).

omitted by further tuning of PD gains and damping. However at least for SI-DP the oscillation is partially inherent to the method since a repulsing potential from the coordinate limits is used.

6   Conclusion

The presented algorithm corresponds to a combination of differential and classical dynamic programming and is applicable to trajectory optimization problems of dynamic systems with continuous state and stage variables and rest-to-rest state transitions. The performance measure in intermediate steps should have a correlation with the global optimization objective. This can be the case by globally optimizing the average values of local objectives and in the presence of high penalties for constraints. An example of a robot trajectory optimization is given, however similar problems may arise in other disciplines as well. The results show the improved performance compared to discrete DP or local optimization at acceptable computation times for offline trajectory planning. The approach is not restricted to second-order trajectories or task redundancy within robotics. Optimizing multiple redundant degrees of freedom is also possible, but without the performance map visualization used throughout the paper. Instead of nullspace optimization, also other local optimization techniques could be used, which leads to a still open future investigation on conditions for the analogy of the proposed method to existing ones like DDP.

Acknowledgements. The author acknowledges the support by the Deutsche Forschungsgemeinschaft (DFG) under grant number 341489206. MATLAB Code to reproduce the results is available at GitHub under free license at https://github.com/SchapplM/robotics-paper_icinco2021.

References

1. Agarwal, A., Nasa, C., Bandyopadhyay, S.: Dynamic singularity avoidance for parallel manipulators using a task-priority based control scheme. Mech. Mach. Theory **96**, 107–126 (2016). https://doi.org/10.1016/j.mechmachtheory.2015.07.013
2. Corinaldi, D., Angeles, J., Callegari, M.: Posture optimization of a functionally redundant parallel robot. In: Lenarčič, J., Merlet, J.-P. (eds.) Advances in Robot Kinematics 2016. SPAR, vol. 4, pp. 101–108. Springer, Cham (2018). https://doi.org/10.1007/978-3-319-56802-7_11
3. De Luca, A., Oriolo, G., Siciliano, B.: Robot redundancy resolution at the acceleration level. Lab. Rob. Autom. **4**, 97–97 (1992)
4. Ferrentino, E., Salvioli, F., Chiacchio, P.: Globally optimal redundancy resolution with dynamic programming for robot planning: a ROS implementation. MDPI Robot. **10**(1), 42 (2021). https://doi.org/10.3390/robotics10010042
5. Gao, J., Pashkevich, A., Caro, S.: Optimization of the robot and positioner motion in a redundant fiber placement workcell. Mech. Mach. Theory **114**, 170–189 (2017). https://doi.org/10.1016/j.mechmachtheory.2017.04.009
6. Gao, Y., Chen, K., Gao, H., Xiao, P., Wang, L.: Small-angle perturbation method for moving platform orientation to avoid singularity of asymmetrical 3-RRR planner parallel manipulator. J. Braz. Soc. Mech. Sci. Eng. **41**(12), 1–18 (2019). https://doi.org/10.1007/s40430-019-2012-4
7. Gosselin, C., Schreiber, L.T.: Kinematically redundant spatial parallel mechanisms for singularity avoidance and large orientational workspace. IEEE Trans. Rob. **32**(2), 286–300 (2016). https://doi.org/10.1109/tro.2016.2516025
8. Gosselin, C., Schreiber, L.T.: Redundancy in parallel mechanisms: a review. Appl. Mech. Rev. **70**(1) (2018). https://doi.org/10.1115/1.4038931
9. Guigue, A., Ahmadi, M., Hayes, M., Langlois, R., Tang, F.: A dynamic programming approach to redundancy resolution with multiple criteria. In: Proceedings 2007 IEEE International Conference on Robotics and Automation, pp. 1375–1380. IEEE (2007). https://doi.org/10.1109/ROBOT.2007.363176
10. Howell, T.A., Jackson, B.E., Manchester, Z.: ALTRO: a fast solver for constrained trajectory optimization. In: 2019 IEEE/RSJ International Conference on Intelligent Robots and Systems (IROS), pp. 7674–7679 (2019). https://doi.org/10.1109/iros40897.2019.8967788

11. Huo, L., Baron, L.: The joint-limits and singularity avoidance in robotic welding. Ind. Robot Int. J. **35**(5), 456–464 (2008). https://doi.org/10.1108/01439910810893626

12. Kotlarski, J., Do Thanh, T., Heimann, B., Ortmaier, T.: Optimization strategies for additional actuators of kinematically redundant parallel kinematic machines. In: 2010 IEEE International Conference on Robotics and Automation (ICRA), pp. 656–661. IEEE (2010). https://doi.org/10.1109/ROBOT.2010.5509982

13. Lantoine, G., Russell, R.P.: A hybrid differential dynamic programming algorithm for constrained optimal control problems. Part 1: theory. J. Optim. Theory Appl. **154**(2), 382–417 (2012). https://doi.org/10.1007/s10957-012-0039-0

14. Lillo, P.D., Chiaverini, S., Antonelli, G.: Handling robot constraints within a set-based multi-task priority inverse kinematics framework. In: 2019 International Conference on Robotics and Automation (ICRA), pp. 7477–7483 (2019). https://doi.org/10.1109/ICRA.2019.8793625

15. Léger, J., Angeles, J.: Off-line programming of six-axis robots for optimum five-dimensional tasks. Mech. Mach. Theory **100**, 155–169 (2016). https://doi.org/10.1016/j.mechmachtheory.2016.01.015

16. Merlet, J.P., Perng, M.W., Daney, D.: Optimal trajectory planning of a 5-axis machine-tool based on a 6-axis parallel manipulator. In: Lenarčič, J., Stanišić, M.M. (eds.) Advances in Robot Kinematics, pp. 315–322. Springer, Dordrecht (2000). https://doi.org/10.1007/978-94-011-4120-8_33

17. Merlet, J.P.: Parallel Robots, Solid Mechanics and Its Applications, vol. 128, 2nd edn. Springer, Dordrecht (2006). https://doi.org/10.1007/1-4020-4133-0

18. Mousavi, S., Gagnol, V., Bouzgarrou, B.C., Ray, P.: Control of a multi degrees functional redundancies robotic cell for optimization of the machining stability. Procedia CIRP **58**, 269–274 (2017). https://doi.org/10.1016/J.PROCIR.2017.04.004

19. Mousavi, S., Gagnol, V., Bouzgarrou, B.C., Ray, P.: Stability optimization in robotic milling through the control of functional redundancies. Robot. Comput.-Integr. Manuf. **50**, 181–192 (2018). https://doi.org/10.1016/j.rcim.2017.09.004

20. Nakamura, Y., Hanafusa, H.: Optimal redundancy control of robot manipulators. Int. J. Robot. Res. **6**(1), 32–42 (1987). https://doi.org/10.1177/027836498700600103

21. Nakamura, Y., Hanafusa, H., Yoshikawa, T.: Task-priority based redundancy control of robot manipulators. Int. J. Robot. Res. **6**(2), 3–15 (1987)

22. Oen, K.T., Wang, L.C.T.: Optimal dynamic trajectory planning for linearly actuated platform type parallel manipulators having task space redundant degree of freedom. Mech. Mach. Theory **42**(6), 727–750 (2007). https://doi.org/10.1016/j.mechmachtheory.2006.05.006

23. Ozgoren, M.K.: Optimal inverse kinematic solutions for redundant manipulators by using analytical methods to minimize position and velocity measures. J. Mech. Robot. **5**(3), 031009 (2013). https://doi.org/10.1115/1.4024294

24. Reiter, A., Müller, A., Gattringer, H.: On higher order inverse kinematics methods in time-optimal trajectory planning for kinematically redundant manipulators. IEEE Trans. Industr. Inf. **14**(4), 1681–1690 (2018). https://doi.org/10.1109/TII.2018.2792002

25. Reveles, D., Pamanes, G.J.A., Wenger, P.: Trajectory planning of kinematically redundant parallel manipulators by using multiple working modes. Mech. Mach. Theory **98**, 216–230 (2016). https://doi.org/10.1016/j.mechmachtheory.2015.09.011

26. Santos, J.C., da Silva, M.M.: Redundancy resolution of kinematically redundant parallel manipulators via differential dynamic programing. J. Mech. Robot. **9**(4), 041016 (2017). https://doi.org/10.1115/1.4036739

27. Schappler, M.: Simulative Optimierung der Bahnplanung mit mehrfacher Redundanz bei der roboterassistierten Laserosteotomie. Bachelor's thesis, Leibniz Universität Hannover, Institut für Mechatronische Systeme (2013). https://doi.org/10.15488/10214

28. Schappler, M., Ortmaier, T.: Singularity avoidance of task-redundant robots in pointing tasks: on nullspace projection and cardan angles as orientation coordinates. In: Proceedings of the 18th International Conference on Informatics in Control, Automation and Robotics (ICINCO 2021) (2021). https://doi.org/10.5220/0010621103380349

29. Sciavicco, L., Siciliano, B.: Modelling and Control of Robot Manipulators. Springer, London (2012). https://doi.org/10.1007/978-1-4471-0449-0

30. Shaw, D., Chen, Y.S.: Cutting path generation of the Stewart-platform-based milling machine using an end-mill. Int. J. Prod. Res. **39**(7), 1367–1383 (2001). https://doi.org/10.1080/00207540010023529

31. Shin, K., McKay, N.: A dynamic programming approach to trajectory planning of robotic manipulators. IEEE Trans. Autom. Control **31**(6), 491–500 (1986). https://doi.org/10.1109/TAC.1986.1104317

32. Smirnov, V., Plyusnin, V., Mirzaeva, G.: Energy efficient trajectories of industrial machine tools with parallel kinematics. In: 2013 IEEE International Conference on Industrial Technology (ICIT), pp. 1267–1272 (2013). https://doi.org/10.1109/icit.2013.6505855

33. Tassa, Y., Mansard, N., Todorov, E.: Control-limited differential dynamic programming. In: 2014 IEEE International Conference on Robotics and Automation (ICRA), pp. 1168–1175. IEEE (2014). https://doi.org/10.1109/ICRA.2014.6907001

34. Zargarbashi, S., Khan, W., Angeles, J.: Posture optimization in robot-assisted machining operations. Mech. Mach. Theory **51**, 74–86 (2012). https://doi.org/10.1016/j.mechmachtheory.2011.11.017

35. Zhu, W., Qu, W., Cao, L., Yang, D., Ke, Y.: An off-line programming system for robotic drilling in aerospace manufacturing. Int. J. Adv. Manuf. Technol. **68**(9–12), 2535–2545 (2013). https://doi.org/10.1007/s00170-013-4873-5

36. Žlajpah, L.: On orientation control of functional redundant robots. In: 2017 IEEE International Conference on Robotics and Automation (ICRA), pp. 2475–2482. IEEE (2017). https://doi.org/10.1109/ICRA.2017.7989288

Signal Processing, Sensors, Systems Modelling and Control

Output Feedback Reference Tracking and Disturbance Rejection for Constrained Linear Systems Using Invariant Sets

Tiago A. Almeida[1], Ana Theresa F. O. Mancini[2(✉)], and Carlos E. T. Dórea[2]

[1] Computer Institute, Federal University of Alagoas (UFAL), Av. Lourival Melo Mota, 57072-900 Maceió, AL, Brazil
tiagoalves@ic.ufal.br

[2] Department of Computer Engineering and Automation, Federal University of Rio Grande do Norte (UFRN), UFRN - CT - DCA, 59078-900 Natal, RN, Brazil
theresa.fmancini@gmail.com, cetdorea@dca.ufrn.br

Abstract. This work deals with the problem of tracking a constant reference signal for constrained linear discrete-time systems in the presence of bounded additive disturbances. By establishing a disturbance model, we design an observer-based output feedback controller such that reference tracking under constraints is achievable. The plant model is augmented by as many disturbance states as there are measured outputs. Output-Feedback Controlled-Invariant polyhedral sets are used to ensure that state and input constraints are satisfied at all times. Based on the available measurements a suitable control sequence is computed via Linear Programming to enforce the constraints and minimize, one step ahead, a distance from the possible states to the equilibrium point associated to the reference signal. The uncertainty on the state is progressively reduced using information about the contraction of invariant sets associated to both the system states and the estimation error. The results are illustrated by numerical examples.

Keywords: Linear systems · Invariant sets · Output feedback · Constraints

1 Introduction

Set-invariance techniques have been successfully applied to handle constrained control problems in the context of linear time invariant systems. In [1] a comprehensive overview of works in this area is presented. Because control and state constraints are mainly associated to polyhedral sets, a possible solution to the regulation problem under linear constraints is to construct a *controlled-invariant* polyhedron contained in the set defined by the state constraints [2,3]. In this case, there exists a corresponding state feedback control such that for any initial condition belonging to the controlled-invariant

This study was financed in part by the Coordenação de Aperfeiçoamento de Pessoal de Nível Superior - Brazil (CAPES) - Finance Code 001, and by the National Council for Scientific and Technological Development - Brazil (CNPq), grant #309862/2019-1.

O. Gusikhin et al. (Eds.): ICINCO 2021, LNEE 1006, pp. 135–150, 2023.
https://doi.org/10.1007/978-3-031-26474-0_7

set the state is driven asymptotically to the origin without constraints violation. However, most known techniques assume that the state is fully measured, which may not be always possible in practice. In this case, we have to resort to output feedback.

In [4] an output feedback structure was studied and conditions were established to evaluate if a given controlled-invariant polyhedral set is *Output-Feedback Controlled-Invariant* (OFCI). OFCI sets are sets wherein the state trajectory can be confined through a suitable sequence of control actions, which are taken based on the measured outputs only. Such a control sequence is computed online through the solution of Linear Programming (LP) problems. In [5] the results of [4] were deepened and extended. There, a dynamic controller structure was proposed, for which an OFCI set can be obtained for an augmented system that comprises the system and compensator states. Moreover, an explicit description of the set of *admissible initial states* (states for which constraints violations can be avoided through output feedback control) is provided and methods to enlarge such sets are discussed. This technique has been further improved in [6] by using information given by the solution of the LP problems mentioned above.

To solve constrained reference tracking control problems with disturbance rejection, Model Predictive Control (MPC) techniques have been used. In [7, 8] disturbance models and observer based approaches are proposed and conditions are derived to attain reference tracking. In [9, 10] the linear system model is augmented with a disturbance model which is used for estimation. Then, MPC is designed to reject the estimated disturbances and track the reference. In [11] the control input is calculated using nominal predictions and the notion of tube of trajectories is used to deal with persistent disturbances and to steer the nominal predicted trajectory to any admissible target. A stabilizing feedback gain can be chosen to reduce the effect of the disturbances on the closed-loop system, but it may possibly affect the stabilizing conditions of the controller. As a consequence, the approach requires the amplitudes of the disturbances to be small.

In this work, we consider output feedback control applied to constant reference tracking with disturbance rejection problem in discrete-time linear systems subject to state and control constraints, and unknown-but-bounded disturbances. The system is assumed to be square, i.e., there are as many inputs as outputs, and subject to piecewise constant disturbances. The system model is augmented with a disturbance model which is used to estimate it and then compensate for its effects. Based on the available measurements and OFCI polyhedral sets a suitable control sequence is computed via Linear Programming which is able to achieve reference tracking under state and control constraints in spite of the disturbances. The uncertainty on the state is progressively reduced using information both from the contraction of an invariant set defined in the estimation error space and from the solution of the LP problem employed to compute the previous control action. The applicability of the proposed controller is illustrated by means of numerical examples.

Notation: $\mathbb{N} = \{0, 1, 2, \ldots\}$. $\overline{1}$ and $\overline{0}$ represent vectors (or matrices) of appropriate dimensions whose components are all equal to 1 and 0, respectively. I represents the identity matrix of appropriate dimension. M_i denotes the ith row of matrix M. If $x \in \mathbb{R}^n$, x_i denotes the ith component of x. A C-set is a convex and compact (closed and bounded) set containing the origin. For a given matrix $M \in \mathbb{R}^{m \times n}$ and a set $X \subset \mathbb{R}^n$,

$MX = \{Mx : x \in X\}$. For any real $\alpha \geq 0$, $\alpha X = \{\bar{x} : \bar{x} = \alpha x, x \in X\}$. For two given sets $X \subset \mathbb{R}^n$ and $Y \subset \mathbb{R}^n$ the symbol $\oplus$ denotes the Minkowski sum operator $X \oplus Y = \{x + y : x \in X, y \in Y\}$. The closed ball with center at x_0 and radius $\varepsilon > 0$ is defined by: $\mathcal{B}(x_0, \varepsilon) = \{x : -\varepsilon \leq x - x_0 \leq \varepsilon\}$. $\{f(k)\}$ denotes a sequence of values of f at $k = 1, 2, \ldots$.

2 Invariant Sets

Consider the square, linear, time-invariant, discrete-time system model, described by:

$$x(k + 1) = Ax(k) + Bu(k) + Ed(k), \tag{1}$$
$$y(k) = Cx(k), \tag{2}$$

where $k \in \mathbb{N}$, $x \in \mathbb{R}^n$ is the state vector, which is assumed not directly measurable, $u \in \mathbb{R}^m$ is the control input, $d \in \mathbb{R}^{n_d}$ is the disturbance signal acting on the system, and $y \in \mathbb{R}^m$ is the measured output. Moreover, the system is subject to control and state constraints: $u \in \mathfrak{U}$ and $x \in \Omega \subseteq \Omega_x$, where $\mathfrak{U} \subset \mathbb{R}^m$ and $\Omega \subset \mathbb{R}^n$ are also C-sets. The pairs (A, B) and (A, C) are assumed to be controllable and observable, respectively.

The constraints on the state variables and control input, and the bounds on the disturbances are given by the following polyhedral sets containing the origin:

$$\Omega_x = \{x : G_x x \leq \bar{1}\}, \quad \mathfrak{U} = \{u : Uu \leq \bar{1}\}, \quad \mathfrak{D} = \{d : Dd \leq \bar{1}\}, \tag{3}$$

where $G_x \in \mathbb{R}^{g_x \times n}$, $U \in \mathbb{R}^{v \times m}$, and $D \in \mathbb{R}^{s \times n_d}$.

We now introduce some basic definitions and notions related to invariant sets.

Definition 1 (Controlled Invariance [2]). *The set $\Omega \subseteq \Omega_x$ is said to be controlled-invariant with contraction rate λ, $0 \leq \lambda < 1$, w.r.t. system (1) if $\forall x \in \Omega$, $\exists u \in \mathfrak{U}$ such that $Ax + Bu + Ed \in \lambda\Omega$, $\forall d \in \mathfrak{D}$.*

If Ω is controlled-invariant with contraction rate λ, then, $\forall x(k) \in \Omega$, there exists a state feedback law $u(x(k))$ such that $x(k + 1) \in \lambda\Omega$, for all admissible disturbances $d \in \mathfrak{D}$. Then, if $x(0) \in \Omega \subseteq \Omega_x$, $\exists u \in \mathfrak{U}$ such that $x(k) \in \Omega$, $\forall k \geq 0$. The condition $x \in \Omega \subseteq \Omega_x$ means that x is required to belong to a controlled-invariant set contained in the set defining the state constraints Ω_x.

The set Ω is the desired region we aim to keep the state vector x limited in. The problem lies on how to accomplish that through the measurements $\{y(k)\}$, $k \in \mathbb{N}$. Consider the following set of *admissible outputs* associated to Ω:

$$\mathcal{Y}(\Omega) = \{y : y = Cx \quad \text{for} \quad x \in \Omega\} = C\Omega. \tag{4}$$

The set $\mathcal{Y}(\Omega) \subset \mathbb{R}^m$ represents all values of y that can be associated to $x \in \Omega$. Thus, if $x \in \Omega$, then $y \in \mathcal{Y}(\Omega)$.

Consider also the following set:

$$\mathfrak{C}(y) = \{x : Cx = y\}. \tag{5}$$

$\mathfrak{C}(y)$ represents the set of states compatible with a *single* measurement $y(k)$.

Hence, from the above settings, set-invariance under output-feedback can be characterized by the next definition:

Definition 2 (OFCI set [4,5]). *The set $\Omega \subseteq \Omega_x$ is said to be Output-Feedback Controlled-Invariant (OFCI) with contraction rate λ, $0 \leq \lambda < 1$, w.r.t. system (1)–(2) if $\forall y \in \mathcal{Y}(\Omega)$, $\exists u \in \mathfrak{U}$ such that $Ax + Bu + Ed \in \lambda\Omega$, $\forall d \in \mathfrak{D}$ and $\forall x \in \Omega$ such that $Cx = y$.*

When Ω is OFCI with contraction rate λ, if $x(k) \in \Omega \subseteq \Omega_x$, then there exists a control $u(y(k)) \in \mathfrak{U}$, computed from the measured output at time k, such that $x(k+1) \in \lambda\Omega$, $\forall k$, in spite of the disturbances $d \in \mathfrak{D}$. [4,5] provide necessary and sufficient conditions to check if a controlled-invariant polyhedral set Ω is OFCI with contraction rate λ, from the solution of LP problems.

Definition 3 (Conditioned Invariance [12]). *The set $\Omega \subseteq \Omega_x$ is said to be conditioned-invariant with contraction rate λ, $0 \leq \lambda < 1$, w.r.t. system (1)–(2) if $\forall y \in \mathcal{Y}(\Omega)$, $\exists v(.)$ such that $Ax + v(.) + Ed \in \lambda\Omega$, $\forall d \in \mathfrak{D}$ and $\forall x \in \Omega$ such that $Cx = y$.*

In the context of state estimation, conditioned-invariant sets are defined on the estimation error space and the variable $v(.)$ is the output injection that limits the estimation error into them. Here, such sets are required to characterize a necessary condition for a given polyhedron, defined on the state space, to be OFCI.

From Definitions 1, 2, and 3, one can conclude what follows [4,5]:

Proposition 1. *Ω is OFCI with contraction rate λ only if it is simultaneously controlled and conditioned-invariant with contraction rate λ.*

3   Disturbance Model for Constant Reference Tracking

It is known [13] that every domain of attraction Ω is also a tracking domain of attraction, i.e. a set of initial states from which one can track asymptotically an admissible reference signal. Therefore, the state-feedback constrained reference tracking problem can be solved once one has found a domain of attraction Ω.

Usually, one focuses only on the bounds of the disturbances. However, bounds can also be provided on their rate of change. Systems with bounded rates of change on exogenous parameters/signals are often referred to as *slowly-varying systems* [14]. The following will be assumed:

Assumption 1. *System (1) is subject to slowly-varying disturbances, characterized by:*

$$d_i(k+1) - d_i(k) = d_{\Delta i}(k) \in \{d_{\Delta i} : |d_{\Delta i}| \leq \eth_i\},$$

where, for $i = 1, \ldots, n_d$, $d_{\Delta i}$ and $\eth_i$ are, respectively, the one-step and the *maximal* (allowable) one-step variations on d_i. In general:

$$d(k+1) - d(k) = d_\Delta(k) \in \mathfrak{D}_\Delta = \{d_\Delta : D_\Delta d_\Delta \leq \overline{1}\}. \tag{6}$$

Then, attaching d as an additional state variable, system (1)–(2) can be written in the extended form:

$$\overline{x}(k+1) = \overline{A}\overline{x}(k) + \overline{B}u(k) + \overline{E}d_\Delta(k), \tag{7}$$

$$y(k) = \overline{C}\overline{x}(k). \tag{8}$$

where $\overline{x}(k) = \begin{bmatrix} x(k) \\ d(k) \end{bmatrix}$, $\overline{A} = \begin{bmatrix} A & E \\ 0 & I \end{bmatrix}$, $\overline{B} = \begin{bmatrix} B \\ 0 \end{bmatrix}$, $\overline{E} = \begin{bmatrix} 0 \\ I \end{bmatrix}$, and $\overline{C} = \begin{bmatrix} C & 0 \end{bmatrix}$.

Generally, one augments the plant model with the purpose of accounting for the effects of plant-model mismatch and persistent disturbances acting on the plant. Augmenting models by a disturbance with integrator is commonly used in offset-free tracking MPC [7–10].

The control structure we will build later requires the following:

Assumption 2. *The pair $(\overline{C}, \overline{A})$ is observable.*

Assumption 2 holds if and only if [7, 8]:

1. The pair (C, A) is observable;
2. The rank of the matrix $\begin{bmatrix} A - I & E \\ C & 0 \end{bmatrix}$ equals $n + n_d$, i.e. it has full column rank.

Thus, for this rank condition to be satisfied the number of augmented disturbance states needs to be less than or equal to the number of outputs, $n_d \le m$ [10]. Consequently, a factor that influences the feasibility of the tracking controller to be presented is that the condition in Assumption 2 may not be satisfied with an increased number of the lumped disturbances. If $(\overline{A}, \overline{C})$ is not observable, one can make the problem feasible by seeking more output information. That is, the problem may become feasible as more output information is accessible [15]. It will be assumed the case $n_d = m$. Here, system (1) may include multiple disturbances in different channels provided that $n_d = m$. In the case of the number of disturbance states n_d different from the number of measured outputs m, $n_d < m$, [10] provides a particular treatment in order to achieve tracking control.

4 Dynamic Output Feedback Controller

A full order dynamic output feedback compensator structure is presented for which an OFCI polyhedron can be constructed from a pair composed by a controlled-invariant and a conditioned-invariant polyhedron defined in the original state space.

Let us consider the system (1)–(2) and the following full-order, possibly nonlinear, compensator [4, 5]:

$$z(k+1) = v[z(k), y(k)], \quad u(k) = \kappa[z(k), y(k)] \tag{9}$$

Then, system (1)–(2) under the compensator (9) can be represented in an extended (or augmented) state space formulation:

$$\xi(k+1) = \hat{A}\xi(k) + \hat{B}\omega(k) + \hat{E}d(k) \tag{10}$$

$$\zeta(k) = \hat{C}\xi(k) \tag{11}$$

where $\xi = \begin{bmatrix} x \\ z \end{bmatrix}$ is the extended state vector, $\omega = \begin{bmatrix} u \\ v \end{bmatrix}$ is the extended input vector, and

$\zeta = \begin{bmatrix} y \\ z \end{bmatrix}$ is the extended output vector. $\hat{A} = \begin{bmatrix} A & \overline{0} \\ \overline{0} & 0 \end{bmatrix}$, $\hat{B} = \begin{bmatrix} B & \overline{0} \\ \overline{0} & I \end{bmatrix}$, $\hat{E} = \begin{bmatrix} E \\ \overline{0} \end{bmatrix}$, and

$\hat{C} = \begin{bmatrix} C & \overline{0} \\ \overline{0} & I \end{bmatrix}$. Moreover, u and v are functions of the extended output vector $\begin{bmatrix} y \\ z \end{bmatrix}$ as
expressed in (9). One will notice that the variable z, which stands for the compensator
state, also appears as an output variable, to explicitly mean that it is "measurable", and
can, therefore, be used for feedback.

Consider now a pair of compact convex polyhedral sets $(\mathcal{V}, \mathcal{S})$, represented by $\mathcal{V} = \{x : G_v x \leq \overline{1}\}$, $\mathcal{S} = \{x : G_s x \leq \overline{1}\}$ and satisfying the following assumptions:

Assumption 3. $\mathcal{S} \subset \mathcal{V} \subset \Omega_x$, $\mathcal{S}$ *is conditioned-invariant* λ_s-*contractive,* $\mathcal{V}$ *is
controlled-invariant* λ_v-*contractive, and* Ω_x *is the set of state constraints.*

We now define a set in the augmented state space which can be proved to be OFCI
candidate according to Proposition 1 [4,5].

Proposition 2. *The polyhedral set:*

$$\hat{\Omega} = \left\{ \xi : \hat{G}\xi \leq \overline{1} \right\} \tag{12}$$

where $\hat{G} = \begin{bmatrix} G_v & \overline{0} \\ G_s & -G_s \end{bmatrix}$, $G_v \in \mathbb{R}^{g_v \times n}$, $G_s \in \mathbb{R}^{g_s \times n}$, *and* $\hat{G} \in \mathbb{R}^{(g_v + g_s) \times 2n}$, *is simultaneously controlled and conditioned-invariant w.r.t. system (10)–(11).*

If $\hat{\Omega}$ is OFCI the constraints (12) can be enforced by a suitable choice of
$(u(k), v(k))$ as functions of $(y(k), z(k))$. From (12), if we interpret the compensator
state $z(k)$ as an estimate of the system state $x(k)$, then $G_s[x(k) - z(k)] \leq \overline{1} \forall k$, i.e.,
the estimation error is bounded by the conditioned-invariant set $\mathcal{S}$.

The compensator structure (9) is quite general, allowing the design of nonlinear
observers. However, as discussed in [5], there is no evidence that a nonlinear observer
would perform better than a linear one. Thus we propose a design method which
employs a Luenberger observer.

4.1 Linear State Estimator

The observer is based on (7)–(8). An estimate $\overline{z}(k) = \begin{bmatrix} \hat{x}^T(k) & \hat{d}^T(k) \end{bmatrix}^T$ of the augmented state $\overline{x}(k) = \begin{bmatrix} x^T(k) & d^T(k) \end{bmatrix}^T$ can be obtained by means of the following
full-order linear Luenberger observer:

$$\overline{z}(k+1) = \overline{A}\overline{z}(k) + \overline{B}u(k) + L[y(k) - \hat{y}(k)] \tag{13}$$

where $\hat{y}(k) = \overline{C}\overline{z}(k)$ is the estimated output and the observer gain $L = \begin{bmatrix} L_x \\ L_d \end{bmatrix}$ is a
parameter that needs to be designed so that the eigenvalues of $(\overline{A} - L\overline{C})$ lie inside the

complex unit disc. L_x and L_d are, respectively, the partitioned gains of L associated with the system and disturbance states.

The dynamics of the estimation error $e(k) = \begin{bmatrix} e_x(k) \\ e_d(k) \end{bmatrix} = \begin{bmatrix} x(k) - \hat{x}(k) \\ d(k) - \hat{d}(k) \end{bmatrix} = \overline{x}(k) - $
$\overline{z}(k)$ is given by:

$$e(k+1) = A_e e(k) + \overline{E} d_\Delta(k), \tag{14}$$

where $A_e = \overline{A} - L\overline{C} = \begin{bmatrix} A - L_x C & E \\ -L_d C & I \end{bmatrix}$. A conditioned invariant polyhedron is required to allow the construction of an OFCI polyhedron with the structure (12). A natural choice is the *minimal Robust Positively Invariant (mRPI) set* [16], which is the smallest invariant set of (14). Let $\overline{S}_m = \left\{ e = \begin{bmatrix} e_x \\ e_d \end{bmatrix} : \overline{G}_s e \le \overline{1} \right\} \subset \mathbb{R}^{n+n_d}$ be the mRPI set of (14) with contraction rate λ_m ($0 \le \lambda_m < 1$). Consider the following (proof given in the Appendix):

Proposition 3. *Let*

$$S_{e_x} = Proj_{e_x} \overline{S}_m, \tag{15}$$

be the projection of $\overline{S}_m$ onto the plant state estimation error space. Then, S_{e_x} is conditioned-invariant w.r.t. (1)–(2).

Proposition 3 makes it possible to consider the dynamics of the system state x and its estimate $\hat{x}$ to perform the OFCI test on the candidate pair $(\mathcal{V}, S_{e_x})$ (see Proposition 2) w.r.t. augmented system (10)–(11). A first choice for $\mathcal{V}$ would be the maximal controlled-invariant set w.r.t. system (1) contained in Ω_x [3].

Given an OFCI pair $(\mathcal{V}, S_{e_x})$, we simply scale S_{e_x} as much as possible up to $\alpha_{e_x} S_{e_x}$ with the pair $(\mathcal{V}, \alpha_{e_x} S_{e_x})$ OFCI with the purpose of enlarging the set of admissible initial states. One can notice that, by doing so, $\alpha_{e_x} S_{e_x}$ may not be included in $\mathcal{V}$ any longer. Consider now the next result (proof given in the Appendix):

Proposition 4. *Let $R[P, \overline{1}] = \{p : Pp \le \overline{1}\} \subset \mathbb{R}^h$, $P \in \mathbb{R}^{n_p \times h}$, be a polyhedral set. Let $\mathfrak{Q} = Proj_q R[P, \overline{1}] \subset \mathbb{R}^n$, $n \le h$. Then, $Proj_q R[P, \alpha \overline{1}] = \alpha(Proj_q R[P, \overline{1}]) = \alpha \mathfrak{Q}$.*

Note that scaling $R[P, \overline{1}]$ by a factor $\alpha > 1$ implies the scaling of $\mathfrak{Q}$ by the same factor.

Proposition 4 allows for the direct scaling of the conditioned-invariant set S_{e_x} by the largest factor $\alpha_{e_x} > 1$ such that the pair $(\mathcal{V}, \alpha_{e_x} S_{e_x})$ remains OFCI rather than scaling the polyhedron $\overline{S}_m$ in the augmented error space and carrying out the projection onto the space e_x.

In [5] a procedure to determine the sets $\overline{\alpha}(k) \overline{S}_m$, $1 < \overline{\alpha}(k) \le \alpha_{e_x}$, which are visited by $e(k)$ before reaching and staying in $\overline{S}_m$ is described. By starting from $\overline{\alpha}(0) = \alpha_m$, it is possible to compute offline a strictly decreasing sequence $\{\overline{\alpha}(k)\}$, $k = 0, 1, \cdots, \overline{k}_m$, for which $e(k) \in \overline{\alpha}(k) \overline{S}_m$ for $k < \overline{k}_m$ and $e(k) \in \overline{S}_m$ for $k \ge \overline{k}_m$, where $\overline{k}_m$ is the smallest value of k such that $\overline{\alpha}(k) \le 1$. This information is used in the control action computation in order to progressively reduce the uncertainty on the system state $x(k)$ and disturbance $d(k)$.

4.2   Tracking Target Calculation

We require a system with the property that for a given constant reference r a unique state-input equilibrium pair (x_{ss}, u_{ss}) in steady-state can be determined. Such a pair must have the reference r as the corresponding equilibrium output, i.e. (x_{ss}, u_{ss}) is the pair state-input for which $y_{ss} = \lim_{k \to \infty} y(k) = r$.

Assumption 4. *The following square matrix is nonsingular:*

$$M_{ss} = \begin{bmatrix} A - I & B \\ C & 0 \end{bmatrix}.$$

The pair (x_{ss}, u_{ss}) is uniquely determined from the solution of the steady-state equation:

$$M_{ss} \begin{bmatrix} x_{ss} \\ u_{ss} \end{bmatrix} = \begin{bmatrix} -E\hat{d}_{ss} \\ r \end{bmatrix}, \tag{16}$$

where $\hat{d}_{ss}$ is the steady-state disturbance estimated by the observer. Assumption 4 implies that the target calculation has a unique solution for all $\hat{d}_{ss}$ and r. The desired set-point r is supposed to be such that the corresponding input u_{ss} and state x_{ss} do not violate the constraints $u_{ss} \in \mathfrak{U}$ and $x_{ss} \in V$ for a given $\hat{d}_{ss}$. A reference signal satisfying the above requirement will be called admissible.

From the steady-state equation (16) one can notice that, at each instant k, one can have a different solution for the pair (x_{ss}, u_{ss}) according to the current estimate $\hat{d}(k)$ of the constant disturbance d_{ss} so that the state x_{ss} and input u_{ss} targets, which should steer the controlled variables to their set-points, are achieved in the observer steady-state. Since one cannot guarantee $\hat{x}(k) \in V$ and $\hat{d}(k) \in \mathfrak{D}$, $\forall k \geq 0$, this implies that during the transient period (for some k) the computed x_{ss} and u_{ss} may not satisfy state and control constraints. However, once $\hat{x}(k)$ and $\hat{d}(k)$ tend, respectively, to a possible value of $x \in V$ and $d \in \mathfrak{D}$, x_{ss} and u_{ss} will also tend to admissible values, as long as the reference is admissible as well.

5   Tracking Controller Design

The overall goal is to obtain a solution to the problem of constant reference tracking under constraints subject to constant disturbances, i.e. $\forall x(0) \in V$, compute $u(k) \in \mathfrak{U}$, $k = 0, 1, \ldots$, based on the measurements $y(k)$ and the estimated augmented state $\bar{z}(k)$ such that $x(k) \in V$, $\forall k \geq 0$, and $\lim_{k \to \infty} y(k) = r$.

5.1   Control Action Calculation

As long as the pair $(V, \alpha_{e_x} S_{e_x})$ forms an OFCI polyhedron, one can proceed with the computation of the control action $u(k)$ that meets the control goal. To this end, the strategy of optimizing in one-step the state trajectory with respect to a closed ball $\mathcal{B}(x_{ss}, \varepsilon)$ is considered, i.e. to steer the state vector x to the smallest ball $\mathcal{B}(x_{ss}, \varepsilon)$

around the steady-state x_{ss} given by $\mathcal{B}(x_{ss}, \varepsilon) = \{x : Hx \leq \varepsilon\overline{1} + Hx_{ss}\}$, where $H = \begin{bmatrix} I & -I \end{bmatrix}^{T}$. Then, one must compute $u \in \mathfrak{U}$ such that $Hx(k+1) \leq \varepsilon\overline{1} + Hx_{ss}$:

$$H(Ax + Bu + Ed) \leq \varepsilon\overline{1} + Hx_{ss} \tag{17}$$

$$\forall x, d : \begin{cases} G_v x \leq \overline{1}, & Cx = y(k) \\ Dd \leq \overline{1}, & \overline{G}_s[\overline{x} - \overline{z}(k)] \leq \overline{\alpha}(k)\overline{1} \end{cases},$$

where $\overline{\alpha}(k)$, $k = 0, \ldots, \overline{k}_m$, with $\overline{\alpha}(0) = \alpha_{e_x}$ and $\overline{\alpha}(k) = 1$ for $k \geq \overline{k}_m$, is the strictly decreasing sequence.

Let the components of vector $\varphi(k) \in \mathbb{R}^{2n}$ be given by the solution of the following LP problem:

$$\varphi_j(k) = \max_{x,d} H_j \begin{bmatrix} A & E \end{bmatrix} \begin{bmatrix} x \\ d \end{bmatrix}, \quad j = 1, \ldots, 2n \tag{18}$$

$$\text{s.t.} \begin{cases} G_v x \leq \overline{1}, & Cx = y(k) \\ Dd \leq \overline{1}, & \overline{G}_s \begin{bmatrix} x - \hat{x}(k) \\ d - \hat{d}(k) \end{bmatrix} \leq \overline{\alpha}(k)\overline{1} \end{cases}.$$

Condition (17) is then equivalent to:

$$HBu - \varepsilon\overline{1} \leq Hx_{ss} - \varphi(k). \tag{19}$$

State and control constraints, that is, $x \in \mathcal{V}$ and $u \in \mathfrak{U}$, can be imposed under output feedback by enforcing the following set of inequalities:

$$G_v Bu \leq \overline{1} - \phi(k) \quad \text{and} \quad Uu \leq \overline{1}, \tag{20}$$

where:

$$\phi_j(k) = \max_{x,d} G_{v_j} \begin{bmatrix} A & E \end{bmatrix} \begin{bmatrix} x \\ d \end{bmatrix}, \quad j = 1, \ldots, g_v \tag{21}$$

$$\text{s.t.} \begin{cases} G_v x \leq \overline{1}, & Cx = y(k) \\ Dd \leq \overline{1}, & \overline{G}_s \begin{bmatrix} x - \hat{x}(k) \\ d - \hat{d}(k) \end{bmatrix} \leq \overline{\alpha}(k)\overline{1} \end{cases}.$$

Although d is bounded, its value acting on the system is unknown and, consequently, a control which enforces the constraints for any $d_{ss} \in \mathfrak{D}$ must be found. Likewise, as the state is not available, the same input u must work for all $x \in \mathcal{V}$ consistent with the measured output y. These two conditions imposed to u can be achieved by calculating the worst case $\begin{bmatrix} x^T & d^T \end{bmatrix}^T$ *row by row* of G_v that may occur. This is the reason for computing the vector $\phi(k)$ in (21). The computation of $\varphi(k)$ in (18) follows the same reasoning used to construct the vector $\phi(k)$: one considers the worst case $\begin{bmatrix} x^T & d^T \end{bmatrix}^T$ (*row by row* of H) with respect to the minimization of ε.

Putting conditions (19) and (20) together, it is possible to compute the control action $u(k)$ such that the system tracks the reference r satisfying simultaneously state and control constraints. Such an action can then be computed *online* from the solution of the following LP:

$$u(k) = \arg \min_{u,\varepsilon} \varepsilon \qquad (22)$$

$$\text{s.t.} \begin{bmatrix} G_v B & \overline{0} \\ U & \overline{0} \\ HB & -\overline{1} \end{bmatrix} \begin{bmatrix} u \\ \varepsilon \end{bmatrix} \leq \begin{bmatrix} \overline{1} - \phi(k) \\ \overline{1} \\ Hx_{ss}(k) - \varphi(k) \end{bmatrix}$$

The first two constraints in (22) guarantee invariance of $\mathcal{V}$ under output feedback. The third constraint accounts for the optimization goal: drive the states consistent with the measurement in one-step to the smallest closed ball around the steady-state x_{ss}. One should notice the dependence on k of x_{ss} in the right-hand side of the third constraint in (22) to exactly mean that x_{ss} is variable (see (16)). Here, it is important to clarify that if the reference r is not admissible the steady-state equation (16) will provide a final solution $x_{ss} \notin \mathcal{V}$ and/or $u_{ss} \notin \mathfrak{U}$. In this case, it is not possible to obtain offset-free tracking, but constraints are always guaranteed to be satisfied.

5.2 Improved Tracking Performance

The difficulty in optimizing performance via output feedback under constraints lies in the fact that a single control action must cope with constraint satisfaction of a set of states consistent with the measurement. The optimization strategy described in the previous section minimizes one step ahead the worst case distance from the set of states consistent with the measurement to the origin. Here, we extend to the reference tracking problem the improvement proposed in [6] for the regulation problem, i.e., to use the solution of the optimization problem in order to further reduce the set of possible states and, as a consequence, improve the convergence of the states to a smaller ball around the equilibrium point.

Let $\epsilon(k)$ be the optimal cost of the LP problem (22). Then, from (17), one can see that, for $k \geq 1$:

$$x(k+1) \in \mathcal{B}(x_{ss}(k), \epsilon(k)) \qquad (23)$$

This information can now be added to the computation of the vectors $\phi(k)$ and $\varphi(k)$ to further reduce the uncertainty on $x(k)$, as follows:

$$\begin{cases} \phi_j(k) = \max_{x,d} G_{vj} \begin{bmatrix} A & E \end{bmatrix} \begin{bmatrix} x \\ d \end{bmatrix}, & j = 1, \ldots, g_v \\ \varphi_j(k) = \max_{x,d} H_j \begin{bmatrix} A & E \end{bmatrix} \begin{bmatrix} x \\ d \end{bmatrix}, & j = 1, \ldots, 2n \end{cases} \qquad (24)$$

$$\text{s.t.} \begin{cases} G_v x \leq \overline{1}, & Cx = y(k), \\ Dd \leq \overline{1}, & G_s \begin{bmatrix} x - \hat{x}(k) \\ d - \hat{d}(k) \end{bmatrix} \leq \overline{\alpha}(k)\overline{1}, . \\ Hx \leq \epsilon(k-1)\overline{1} + Hx_{ss}(k) \end{cases}$$

This way, our proposed optimization problem guarantees that the set of states $x(k)$ consistent with the measured output $y(k)$ belongs to both the controlled-invariant set, by forcing $G_v x \leq \bar{1}$, and the closed ball $\mathcal{B}(x_{ss}(k-1), \epsilon(k-1))$. As a result, the set of states consistent with the measurements becomes smaller improving, therefore, the performance of the tracking controller.

It is worth mentioning that, as shown in [5], we are able to compute offline the decreasing sequence $\{\bar{\alpha}(k)\}$, which defines the contraction rate of the invariant sets related to the estimation error, based on λ_m. On the other hand, we do not have the same previous information for the state $x(k)$. We have to compute $\epsilon(k)$ online because the closed ball $\mathcal{B}(x_{ss}, \epsilon)$ is not an invariant set, then, there is no defined contraction rate.

6 Numerical Example

Example 1. *Consider the discrete-time system (1)–(2) for which:*

$$A = \begin{bmatrix} 0.9347 & 0.5194 \\ 0.3835 & 0.8310 \end{bmatrix}, B = \begin{bmatrix} -1.4462 \\ -0.7012 \end{bmatrix}, E = \begin{bmatrix} 1 \\ 0 \end{bmatrix}, \text{ and } C = \begin{bmatrix} 0.5 & 0.5 \end{bmatrix}.$$

State and control constraints are given respectively by:

$$\Omega_x = \{x : |x_i| \leq 4, i = 1, 2.\} \quad \text{and} \quad \mathfrak{U} = \{u : |u| \leq 1\}.$$

Bounds for disturbance and its maximal one-step change are given respectively by:

$$\mathfrak{D} = \{d : |d| \leq 1\} \quad \text{and} \quad \mathfrak{D}_\Delta = \{d_\Delta : |d_\Delta| \leq 0.1\}.$$

A λ_v-contractive controlled-invariant set $\mathcal{V}$ contained in Ω_x with a contraction rate of $\lambda_v = 0.99$ was computed by using the algorithm proposed in [3]. It can readily be verified that Assumptions 2 and 4 do hold.

One can then proceed with the design of a dynamic compensator by computing a λ_m-contractive conditioned-invariant set $\overline{\mathcal{S}}_m$ w.r.t. system (14) under constraints (6), with $\lambda_m = 0.5$. The gain $L = \begin{bmatrix} L_x{}^T | L_d \end{bmatrix}^T = \begin{bmatrix} 3.2779 & 0.6535 | 1.3032 \end{bmatrix}^T$ was designed to result in the eigenvalues of $A_e = \overline{A} - L\overline{C}$ at $0, 0.4$, and 0.4. By using the conditions proposed in [4], it turns out that the pair $(\mathcal{V}, \mathcal{S}_{e_x} = Proj_{e_x} \overline{\mathcal{S}}_m)$ forms an OFCI polyhedron w.r.t. the extended system (10)–(11). In fact, any pair $(\mathcal{V}, \alpha_{e_x} \mathcal{S}_{e_x})$, with $1 \leq \alpha_{e_x} \leq 3.7$ also forms an OFCI polyhedron. The sets $\Omega_x, \mathcal{V}, \mathcal{S}_{e_x}$, and $\alpha_{e_x} \mathcal{S}_{e_x}$ (with $\alpha_{e_x} = 3.7$) are shown in Fig. 1.

As it was assumed a maximal one-step change for d, this means that d can change at most $d_\Delta = \pm 0.1$. The following situation will be considered: initially, the system has to track a constant reference $r = 1$ with a constant disturbance $d = 0.5$; then, after 20 seconds the reference changes to $r = 1.5$; finally, after 35 seconds the disturbance d increments $d_\Delta = 0.1$.

In Fig. 1 it is also possible to see the state vector trajectory under constant disturbance, starting from the initial augmented state $\overline{x}(0) = \begin{bmatrix} x(0)^T | d \end{bmatrix}^T = \begin{bmatrix} 0 & 0 | 0.5 \end{bmatrix}^T$, with $\overline{z}(0) = \begin{bmatrix} \hat{x}(0)^T | \hat{d} \end{bmatrix}^T = \begin{bmatrix} -0.2 & 4 | 4 \end{bmatrix}^T$. The observer was initialized such that $[\overline{x}(0) - \overline{z}(0)] \in \alpha_{e_x} \overline{\mathcal{S}}_m$, which represents the uncertainty on the initial condition. By

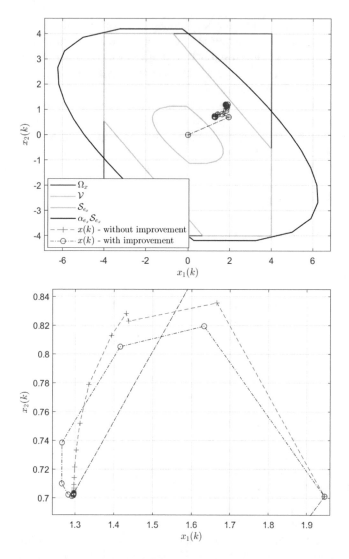

Fig. 1. Set of state constraints Ω_x, the mRPI set $\mathcal{S}_{e_x}$, the pair $(\mathcal{V}, \alpha_{e_x}\mathcal{S}_{e_x})$, and state vector trajectory satisfying the constraints.

zooming in near the steady-state $x_{ss} = \begin{bmatrix} 1.2968 \ 0.7032 \end{bmatrix}^T$ (corresponding to $r = 1$ and $d = 0.5$), we see that the state trajectory resulting from the control action using the additional constraint reaches it faster than the one resulting from the controller without improved tracking performance. The rightmost sample corresponds to $x(1)$.

A sequence $\{\bar{\alpha}(k)\}_{k=0}^{\bar{k}_m}$, with $\bar{\alpha}(0) = \alpha_{e_x}$, was computed such that $e_x(k) \in \bar{\alpha}(k)\mathcal{S}_{e_x}$ for $k < \bar{k}_m$ and $e_x(k) \in \mathcal{S}_{e_x}$ for $k \geq \bar{k}_m$. For this example, the sequence is given by $\{\bar{\alpha}(k)\}_{k=0}^{2} = \{3.7, 1.8515, 0.9265\}$. Figure 2 portrays the sets $\alpha_{e_x}\mathcal{S}_{e_x}$ (with $\alpha_{e_x} = 3.7$), $\overline{\alpha}(1)\mathcal{S}_{e_x}$, $\overline{\alpha}(2)\mathcal{S}_{e_x}$, and $\mathcal{S}_{e_x}$ and the error vector trajectory $e_x(k)$.

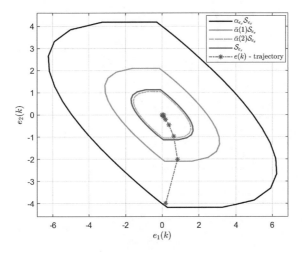

Fig. 2. Polyhedral sets $\alpha_{e_x}\mathcal{S}_{e_x}$, $\overline{\alpha}(1)\mathcal{S}_{e_x}$, $\overline{\alpha}(2)\mathcal{S}_{e_x}$, and $\mathcal{S}_{e_x}$ and error vector trajectory satisfying the constraints.

In Fig. 3 it is possible to see that when considering the additional constraint the state trajectory is associated to a sequence with smaller values of $\varepsilon(k)$. It is worth mentioning that when the system is not affected by disturbances, then $\varepsilon(k)$ tends to 0.

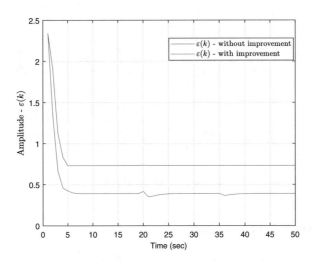

Fig. 3. Evaluation of $\varepsilon(k)$.

In Figs. 4 and 5 are shown respectively the corresponding time evolution of the output $y(k)$ and the control signal $u(k)$. It is possible to see how the effect of the constant disturbance is removed using the control action (22) and hence achieving offset-free tracking.

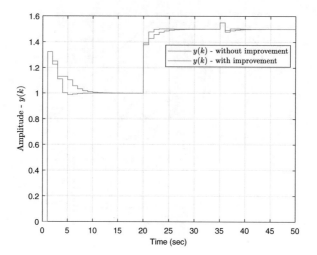

Fig. 4. Time response $y(k)$.

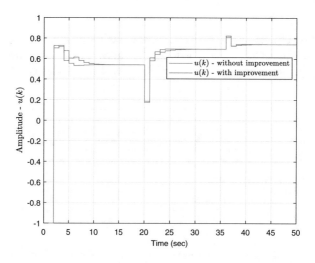

Fig. 5. Control signal $u(k)$.

7   Conclusions and Future Work

In this work, the design of a dynamic output feedback controller for tracking of constant reference signals with rejection of constant disturbances for constrained, linear, discrete-time systems, was presented. This problem is often referred to as offset-free tracking control.

Based on the concepts of Output-Feedback Controlled-Invariant (OFCI) sets and set-invariant estimators a suitable control sequence, through the solution of a Linear Programming problem, can be computed online to enforce the constraints and minimize in one-step the distance from the state trajectory to a closed ball around the steady-state target for constant references in the presence of constant disturbances. An augmented system that is composed by the state and the disturbance was considered for the estimator. The augmented system takes into account the maximal one-step change of the disturbances. The set of states consistent with the measurement is further reduced, by taking into account the results of an optimization problem solved in the previous step, leading to an improved convergence of the states to the required equilibrium point.

As future work, some important points are to be explored:

– Establish a criterion to design the output injection gain L;
– Investigate the case where disturbances enter the input of the system.

Appendix

Proof of Proposition 3. From (7) and (13), one has that:

$$x(k+1) = Ax(k) + Ed(k) + Bu(k)$$
$$\hat{x}(k+1) = A\hat{x}(k) + E\hat{d}(k) + Bu(k) + v(y(k) - \hat{y}(k)),$$

where $v(y(k) - \hat{y}(k)) = L_x[y(k) - \hat{y}(k)]$. Then:

$$e_x(k+1) = x(k+1) - \hat{x}(k+1) \tag{25}$$
$$= Ae_x(k) + E[d(k) - \hat{d}(k)] - v(y(k) - \hat{y}(k)).$$

If $e_x(k) \in \mathcal{S}_{e_x}$, then $\exists \hat{d}(k) : \begin{bmatrix} e_x(k) \\ d(k) - \hat{d}(k) \end{bmatrix} \in \overline{\mathcal{S}}_m$. Since $\overline{\mathcal{S}}_m$ is mRPI, then, the designed L is such that $\begin{bmatrix} e_x(k+1) \\ d(k+1) - \hat{d}(k+1) \end{bmatrix} \in \overline{\mathcal{S}}_m$. Hence $e_x(k+1) \in Proj_{e_x} \overline{\mathcal{S}}_m$.

Then, from (25) there exists $\tilde{v}(y(k) - \hat{y}(k)) = -v(y(k) - \hat{y}(k)) - E\hat{d}(k)$ such that $e_x(k+1) = Ae_x(k) + Ed(k) + \tilde{v}[y(k) - \hat{y}(k)] \in Proj_{e_x} \overline{\mathcal{S}}_m$. This proves, from Definition 3, that $Proj_{e_x} \overline{\mathcal{S}}_m$ is conditioned-invariant w.r.t. system (1)–(2).

Proof of Proposition 4. Partition the vector $p \in \mathbb{R}^h$ into $p = \begin{bmatrix} q \\ l \end{bmatrix}$, where $q \in \mathbb{R}^n$ and $l \in \mathbb{R}^{(h-n)}$. Accordingly, partition the matrix $P \in \mathbb{R}^{n_p \times h}$ into $P = \begin{bmatrix} P' & P'' \end{bmatrix}$, where $P' \in \mathbb{R}^{n_p \times n}$ and $P'' \in \mathbb{R}^{n_p \times (h-n)}$. Then:

$$Pp \leq \overline{1} \implies \begin{bmatrix} P' & P'' \end{bmatrix} \begin{bmatrix} q \\ l \end{bmatrix} \leq \overline{1} \implies P'q + P''l \leq \overline{1}.$$

There exists a matrix T [3] whose row vectors form a minimal generating set of the non-negative left kernel of matrix P'' such that:

$$TP'q + \underbrace{TP''}_{\overline{0}}l \leq T\overline{1} \implies TP'q \leq T\overline{1} \implies R[TP', T\overline{1}].$$

Clearly, $R[TP', T\overline{1}] = Proj_q R[P, \overline{1}] = \mathfrak{Q}$.

Considering now the homothetic polyhedron $R[P, \alpha\overline{1}]$, with $\alpha > 1$, and following the same procedure above, the projection of $R[P, \alpha\overline{1}]$ onto the space $\mathbb{R}^n$ results in $[TP', \alpha(T\overline{1})] = \alpha\mathfrak{Q}$. This shows the statement of the Proposition.

References

1. Blanchini, F., Miani, S.: Set-Theoretic Methods in Control, Birkhäuser, Systems & Control: Foundations & Applications, 2nd edn. Birkhäuser Verlag, Basel (2015)
2. Blanchini, F.: Ultimate boundedness control for uncertain discrete-time systems via set-induced Lyapunov functions. IEEE Trans. Autom. Control **39**(2), 428–433 (1994)
3. Dórea, C.E.T., Hennet, J.C.: (A, B)-invariant polyhedral sets of linear discrete-time systems. J. Optim. Theory Appl. **103**(3), 521–542 (1999)
4. C. E. T. Dórea, Output-feedback controlled-invariant polyhedra for constrained linear systems. In: Proceeding of 48h IEEE Conference Decision Control (CDC) and 28th Chinese Control Conference, Shanghai, pp. 5317–5322 (2009)
5. Almeida, T.A., Dórea, C.E.T.: Output feedback constrained regulation of linear systems via controlled-invariant sets. IEEE Trans. Autom. Control **66**(7), 3378–3385 (2021). https://doi.org/10.1109/TAC.2020.3020100
6. Mancini A., Almeida T. and Dórea C.: Improved output feedback control of constrained linear systems using invariant sets. In: Proceedings of 18th International Conference on Informatics in Control, Automation and Robotics, vol. 1, ICINCO, pp. 566–573 (2021). https://doi.org/10.5220/0010556705660573. (ISBN 978-989-758-522-7)
7. Muske, K.R., Badgwell, T.A.: Disturbance modeling for offset-free linear model predictive control. J. Process Control **12**(5), 617–632 (2002)
8. Pannocchia, G., Rawlings, J.B.: Disturbance models for offset-free model-predictive control. AIChE J. **49**(2), 426–437 (2003)
9. Maeder, U., Morari, M.: Offset-free reference tracking for predictive controllers. In: 46th IEEE Conference on Decision and Control, pp. 5252–5257 (2007)
10. Maeder, U., Borrelli, F., Morari, M.: Linear offset-free model predictive control. Automatica **45**(10), 2214–2222 (2009)
11. Limon, D., Alvarado, I., Alamo, T., Camacho, E.F.: Robust tube-based MPC for tracking of constrained linear systems with additive disturbances. J. Process Control **20**(3), 248–260 (2010)
12. Dórea, C.E.T., Pimenta, A.C.C.: Design of set-invariant estimators for linear discrete-time systems. In: Proceedings of the 44th IEEE Conference on Decision and Control, Seville, Spain, pp. 7235–7240 (2005)
13. Blanchini, F., Miani, S.: Any domain of attraction for a linear constrained system is a tracking domain of attraction. SIAM J. Control. Optim. **38**(3), 971–994 (2000)
14. Nilsson, M., Klintberg, E., Rumschinski, P., Mardh, L.J.: Admissible sets for slowly-varying discrete-time systems. Automatica **112**, 108676 (2020)
15. Li, S., Yang, J., Chen, W.H., Chen, X.: Disturbance Observer-Based Control: Methods and Applications. CRC Press, Boca Raton (2014)
16. Rakovic, S.V., Kerrigan, E.C., Kouramas, K.I., Mayne, D.Q.: Invariant approximations of the minimal robust positively invariant set. IEEE Trans. Autom. Control **50**(3), 406–410 (2005)

Output-Feedback Model Predictive Control Using Set of State Estimates

Lenka Kuklišová Pavelková[(✉)] and Květoslav Belda

The Czech Academy of Sciences, Institute of Information Theory and Automation,
Pod Vodárenskou věží 4, 182 00 Prague 8, Czech Republic
{pavelkov,belda}@utia.cas.cz
https://www.utia.cas.cz

Abstract. The paper deals with an algorithm of output-feedback model predictive control (MPC) where the required point state estimate is selected from the set of possible estimates. The involved state estimator is based on an approximate uniform Bayesian filter. In the paper, there are compared conservative mean and progressive composite state estimates. The proposed method is illustrated by the motion control of a specific robotic system.

Keywords: Output-feedback control · Model predictive control · State estimation · Bayesian methods · Robotic system · Bounded disturbances

1 Introduction

The output-feedback model predictive control (MPC) is popular as the states of the involved state space model are often unmeasurable in the praxis [2]. In such case, the control performance depends on the quality of the state estimates. This quality is usually influenced by uncertainties that are related to a model inaccuracy and to unmeasured noises. The statistics of these uncertainties are rarely known. In many practical applications, they are only known to be bounded, and any additional information about their nature and properties is unavailable [10]. Therefore, the output-feedback MPC, that considers a bounded uncertainty, is one of the recent research concerns.

The estimation techniques to cope with bounded disturbances are based either on stochastic or set-membership approach. Set-membership algorithms provide state estimates that are confined in constrained sets such as boxes [20], zonotopes [23], ellipsoids [16] or their combination [26]. Stochastic state estimation is based on particle filtering [22] or can resemble Kalman filter with data and time update steps [8]. Stochastic and set-membership paradigms are merged in [6].

Set-membership state estimation has been used e.g. in [21], [5] while in [27], a specific robust Kalman filter has been used. Recently, a tube-based robust MPC scheme was proposed where the states are bounded by the tubes whose center is the state of the nominal system [11, 15, 19, 25]. The paper [17] combines set-membership estimation with prediction tubes.

In our research, we focus on the output-feedback MPC intended for industrial stationary robots-manipulators, specifically parallel kinematic machine (PKM) [18] where

© The Author(s), under exclusive license to Springer Nature Switzerland AG 2023
O. Gusikhin et al. (Eds.): ICINCO 2021, LNEE 1006, pp. 151–162, 2023.
https://doi.org/10.1007/978-3-031-26474-0_8

the system outputs correspond to the Cartesian coordinates and angular position. The unmeasured states consist of the relevant velocities. In this setting, measurements are often influenced by physically bounded uncertainties.

We propose an algorithm of output-feedback MPC for discrete-time systems influenced by bounded state and output disturbances. The required estimates are provided by the Bayesian state estimator presented in [8]. The control aim is to follow a given reference trajectory. The paper builds on the previous authors works presented in [12, 13], and proposes a novel way how to choose a point estimate from the admissible set of state estimates. In [12], the center of a set estimate was chosen as a point estimate in each simulation step. In [13], the choice of the point estimate was included into the optimization step. Here, we will choose it in advance based on the previous state evolution.

The paper is organised as follows. This section ends with a summary of the notation used. Section 2 introduces a linear state space model with uniform disturbances including its approximate Bayesian estimation. In Sect. 3, two algorithms of output-feedback MPC using the above mentioned model are explained. Section 4 presents experiments with a model of the parallel kinematic machine where the proposed control scheme is applied to the reference tracking. Section 5 concludes the paper.

Notation. Matrices are in capital letters (e.g. A), vectors and scalars are in lowercase letters (e.g. b). A_{ij} is the element of a matrix A on i-th row and j-th column. A_i denotes the i-th row of A. Column vectors are considered, where z_t denotes the value of a vector variable z at a discrete-time instant $t \in \{1, \cdots, \bar{t}\}$; $z_{t;i}$ is the i-th entry of z_t; $\underline{z}$ and $\overline{z}$ are lower and upper bounds on z, respectively. $\hat{z}$ denotes the estimate of z. The symbol $f(\cdot|\cdot)$ denotes a conditional probability density function (pdf); names of arguments distinguish respective pdfs; no formal distinction is made between a random variable, its realisation and an argument of the pdf. $\mathcal{U}_z(\underline{z}, \overline{z})$ denotes a multivariate uniform distribution of z, $\underline{z} \leq z \leq \overline{z}$, inequalities are meant entrywise; $\|.\|_2^2$ means the squared Euclidean norm.

2   Bayesian State Estimation of LSU Model

A linear state space model with uniform disturbances (LSU model) is defined as

$$x_t = \underbrace{A_t\, x_{t-1} + B_t\, u_{t-1}}_{\tilde{x}_t} + \nu_t,\ \nu_t \sim \mathcal{U}_\nu(-\rho, \rho) \tag{1}$$

$$y_t = \underbrace{C x_t}_{\tilde{y}_t} + n_t, \qquad n_t \sim \mathcal{U}_n(-r, r) \tag{2}$$

where A_t, B_t are time varying model matrices; $C = [\text{I}\ 0]$; $\tilde{x}_t$ and $\tilde{y}_t$ correspond to the nominal values of x_t and y_t, respectively; ν_t and n_t are independent and identically distributed (i.i.d.) state and observation disturbances. They are uniformly distributed within an orthotope with known bounds ρ and r, respectively.

In the Bayesian filtering framework [9], a controlled system is described by the following pdfs:

$$\text{time evolution model:} \quad f\left(x_t | x_{t-1}, u_{t-1}\right) \tag{3}$$

$$\text{observation model:} \quad f\left(y_t | x_t\right) \tag{4}$$

$$\text{prior pdf:} \quad f\left(x_0\right) \tag{5}$$

Bayesian state estimation (filtering) consists in the evolution of the posterior pdf $f(x_t | d(t))$ where $d(t)$ is a sequence of observed data records $d_t = (y_t, u_t)$, $d_0 \equiv u_0$. The evolution of posterior pdf $f(x_t | d(t))$ is described by a two-steps recursion that starts from the prior pdf $f(x_0 | u_0) \equiv f(x_0)$ (5): (i) *time update* that uses theoretical knowledge described by model (3) and reflects the evolution $x_{t-1} \rightarrow x_t$; it provides prediction $f(x_t | d(t-1))$, and (ii) *data update* that uses theoretical knowledge described by model (4) and incorporates information about data d_t; it provides The LSU model (1), (2) can be equivalently described, using pdf notation (3)–(5), as follows

$$f(x_t | u_{t-1}, x_{t-1}) = \mathcal{U}_x(\tilde{x}_t - \rho, \tilde{x}_t + \rho) \tag{6}$$

$$f(y_t | x_t) = \mathcal{U}_y(\tilde{y}_t - r, \tilde{y}_t + r) \tag{7}$$

$$f(x_0) = \mathcal{U}_x(\underline{x}_0, \overline{x}_0) \tag{8}$$

The exact solution of the Bayesian filtering of LSU model (6), (7) leads to a very complex form of posterior pdf. Recently, an approximate Bayesian state estimation was proposed by one of authors [7]. It provides the evolution of the uniformly distributed posterior pdf $f(x_t | d(t))$ as follows.

Time Update – time update starts at $t = 1$ with $\underline{m}_0 = \underline{x}_0$, $\overline{m}_0 = \overline{x}_0$ and holds

$$f(x_t | d(t-1)) \approx \prod_{i=1}^{\ell} \mathcal{U}_{x_{t;i}}(\underline{m}_{t;i} - \rho_i, \overline{m}_{t;i} + \rho_i) = \mathcal{U}_{x_t}(\underline{m}_t - \rho, \overline{m}_t + \rho), \tag{9}$$

where $\underline{m}_t = [\underline{m}_{t;1}, \ldots, \underline{m}_{t;\ell}]^T$, $\overline{m}_t = [\overline{m}_{t;1}, \ldots, \overline{m}_{t;\ell}]^T$, ℓ is the size of x,

$$\underline{m}_{t;i} = \sum_{j=1}^{\ell} \min(A_{ij} \underline{x}_{t-1;j} + B_i u_{t-1}, A_{ij} \overline{x}_{t-1;j} + B_i u_{t-1}), \tag{10}$$

$$\overline{m}_{t;i} = \sum_{j=1}^{\ell} \max(A_{ij} \underline{x}_{t-1;j} + B_i u_{t-1}, A_{ij} \overline{x}_{t-1;j} + B_i u_{t-1}).$$

Data Update – in data update, the observation y_t (7) is processed by the Bayes rule together with the prior (9) from the time update as $y_t - r \le C x_t \le y_t + r$. The resulting uniform pdf posses a support in the form of polytope. It is approximated by a uniform pdf with an orthotopic support

$$f(x_t | d(t)) \approx \mathcal{U}_{x_t}(\underline{x}_t, \overline{x}_t). \tag{11}$$

This approximation is based on a minimising of Kullback-Leibler divergence of two pdfs [7]. The result of (11) says that the estimate $\hat{x}_t$ belong to a set

$$\hat{x}_t \in \langle \underline{x}_t, \overline{x}_t \rangle \tag{12}$$

where all points have the same probability.

For the intended task of the output-feedback MPC, we need a state point estimate. In [12], this point estimate corresponded to the center of a set estimate (12) at each control design step. In [13], we did not choose the particular point estimate to be used in the control design but consider the whole set (12). In each time step, the optimization run several times for a chosen sequence of points from this set. Then, the point connected with a minimal cost was chosen for the control input computation.

Now, we will come back to the concept of a priori chosen estimate. Contrary to the paper [12], the choice will not be fixed to the center of the set estimate but will depend on the shape of reference trajectory and on the previous state evolution. Details concerning the mentioned choice are presented in Sect. 4.

3   Control Design

To design an optimal control action, MPC employs predictions of expected future outputs of controlled system represented by a state space model. The equations of predictions are composed using current state estimate in nominal parts of model (1) and (2). For simplicity, we omit here the time indices, i.e., $A_t \rightarrow A$ and $B_t \rightarrow B$, as for one optimisation step, we will consider the matrices to be constant within a prediction horizon N that is for control horizon as well.

Prediction Equations – *positional control algorithm* [4, 12]:

$$\hat{Y}_{t+1} = \left[\hat{y}_{t+1}^T, \cdots, \hat{y}_{t+N}^T\right]^T = F_1 \hat{x}_t + G_1 U_t,$$

$$U_t = \left[u_t^T, \cdots, u_{t+N-1}^T\right]^T \tag{13}$$

$$F_1 = \begin{bmatrix} CA \\ \vdots \\ CA^{N-1} \\ CA^N \end{bmatrix}, \quad G_1 = \begin{bmatrix} CB & 0 & \cdots & 0 \\ \vdots & & \ddots & \vdots \\ CA^{N-2}B & \cdots & CB & 0 \\ CA^{N-1}B & \cdots & CAB & CB \end{bmatrix}.$$

To achieve integral property in the design, the nominal parts of model (1) and (2) are rewritten in incremental forms as follows [13]

$$\Delta\hat{x}_{t+1} = \hat{x}_{t+1} - \hat{x}_t = A\,\Delta\hat{x}_t + B\,\Delta u_t$$
$$\Delta\hat{y}_{t+1} = \hat{y}_{t+1} - y_t = C\,\Delta\hat{x}_{t+1}. \tag{14}$$

Prediction Equations – *incremental control algorithm*: the equations are composed analogically to the positional algorithm but using the model (14):

$$\Delta\hat{x}_{t+j} = A^j\,\Delta\hat{x}_t + \sum_{i=1}^{j} A^{i-1}B\Delta u_{t+j-i} \tag{15}$$

$$\Delta\hat{y}_{t+j} = CA^j\,\Delta\hat{x}_t + \sum_{i=1}^{j} CA^{i-1}B\Delta u_{t+j-i} \tag{16}$$

The evolution of the full-value predictions of the system outputs $\hat{y}$ is

$$\hat{y}_{t+j} = y_t + \sum_{i=1}^{j} \Delta\hat{y}_{t+i} \tag{17}$$

The relevant matrix notation of (17) is as follows

$$\hat{Y}_{t+1} = [\hat{y}_{t+1}^T \cdots \hat{y}_{t+N}^T]^T = F_\mathrm{I}\, y_t + F_2\, \Delta\hat{x}_t + G_2\, \Delta U_t, \tag{18}$$

$$\Delta U_t = \left[\Delta u_t^T, \cdots, \Delta u_{t+N-1}^T\right]^T \tag{19}$$

$$F_\mathrm{I} = [I \cdots I]^T, \quad F_2 = \begin{bmatrix} CA \\ \vdots \\ \sum_{i=1}^{N} CA^i \end{bmatrix}, \quad G_2 = \begin{bmatrix} CB & \cdots & 0 \\ \vdots & \ddots & \vdots \\ \sum_{i=1}^{N} CA^{i-1}B & \cdots & CB \end{bmatrix}$$

The behaviour of a control process is influenced by the choice of the cost function. MPC involves a quadratic cost function that balances control errors, i.e. differences between predicted outputs and reference values, against amount of input energy given by control vector (for the the positional algorithm) or control increments (for the incremental algorithm).

Cost Function: has the form

$$J_t = (\hat{Y}_{t+1} - W_{t+1})^T Q_{YW}^T Q_{YW} (\hat{Y}_{t+1} - W_{t+1}) + \mathbb{U}_t^T Q_{\mathbb{U}}^T Q_{\mathbb{U}} \mathbb{U}_t \tag{20}$$

where $\mathbb{U}_t = U_t$ for the *positional algorithm* [12] and $\mathbb{U}_t = \Delta U_t$ for the *incremental algorithm* [13]; $W_{t+1} = \left[w_{t+1}^T, \cdots, w_{t+N}^T\right]^T$ represents a sequence of references.

Optimality Criterion: is defined as follows

$$\min_{\mathbb{U}_t} J_t (\hat{Y}_{t+1}, W_{t+1}, \mathbb{U}_t), \ \mathbb{U}_t \in \{U_t, \Delta U_t\} \tag{21}$$

s. t. state space model $(1), (2)$ (or $(15), (16)$)

state estimate $\hat{x}_t$ (12)

where $\hat{Y}_{t+1}$ are prediction Eq. (13) or (18), respectively. The involved cost function J_t (20) is rewritten into the square-root form

$$J_t = \mathbb{J}_t^T \mathbb{J}_t \tag{22}$$

For the *Positional algorithm*, the square-root $\mathbb{J}_t$ of J_t (22) is

$$\mathbb{J}_t = \begin{bmatrix} Q_{YW} & 0 \\ 0 & Q_U \end{bmatrix} \begin{bmatrix} \hat{Y}_{t+1} - W_{t+1} \\ U_t \end{bmatrix} = \begin{bmatrix} Q_{YW}(G_1\, U_t - Z_{pos}) \\ Q_U\, U_t \end{bmatrix} \tag{23}$$

where $Z_{pos} = W_{t+1} - F_\mathrm{I}\, \hat{x}_t$.

For the *Incremental algorithm*, the square-root $\mathbb{J}_t$ of J_t (22) is

$$
\mathbb{J}_t = \begin{bmatrix} Q_{YW} & 0 \\ 0 & Q_{\Delta U} \end{bmatrix} \begin{bmatrix} \hat{Y}_{t+1} - W_{t+1} \\ \Delta U_t \end{bmatrix} = \begin{bmatrix} Q_{YW} \left(G_2 \, \Delta U_t - Z_{inc} \right) \\ Q_{\Delta U} \Delta U_t \end{bmatrix} \tag{24}
$$

where $Z_{inc} = W_{t+1} - F_I \, y_t - F_2 \, \Delta \hat{x}_t$ and Q_{YW}, $Q_{\Delta U}$ and Q_U are penalisation matrices defined as follows

$$
Q_\diamond^T Q_\diamond = \begin{bmatrix} Q_*^T Q_* & & 0 \\ & \ddots & \\ 0 & & Q_*^T Q_* \end{bmatrix} \begin{array}{l} \text{subscripts } \diamond, * : \\ \diamond \in \{YW, \, \Delta U, \, U\} \\ * \in \{yw, \, \Delta u, \, u\} \end{array} \tag{25}
$$

Optimization: consist in the minimisation of the cost function. Considering the square-root $\mathbb{J}_t$ (23) or (24), the minimisation, as a specific solution of least-squares problem, leads to the following algebraic equations for (23) [12]:

$$
\underbrace{\begin{bmatrix} Q_{YW} \, G_1 & Q_{YW} \, Z_{pos} \\ Q_U & 0 \end{bmatrix}}_{\mathcal{A} \qquad b} \begin{bmatrix} U_t \\ -I \end{bmatrix} = 0 \tag{26}
$$

and for (24) [13]:

$$
\underbrace{\begin{bmatrix} Q_{YW} \, G_2 & Q_{YW} \, Z_{inc} \\ Q_{\Delta U} & 0 \end{bmatrix}}_{\mathcal{A} \qquad b} \begin{bmatrix} \Delta U_t \\ -I \end{bmatrix} = 0 \tag{27}
$$

The over-determined system (26) or (27), respectively, can be transformed by orthogonal-triangular decomposition [14] so that matrix $[\, \mathcal{A} \;\; b \,]$ is transformed into upper triangle matrix R_1, and solved for unknown $\mathbb{U}_t \in \{U_t, \Delta U_t\}$. This transformation is indicated by the following equation diagram

$$
\begin{bmatrix} \mathcal{A} & b \end{bmatrix} \begin{bmatrix} \mathbb{U}_t \\ -I \end{bmatrix} = 0 \;\; \Rightarrow \;\; \begin{bmatrix} R_1 & c_1 \\ 0 & c_z \end{bmatrix} \begin{bmatrix} \mathbb{U}_t \\ -I \end{bmatrix} = 0 \tag{28}
$$

where the vector c_z represents a loss vector. Its Euclidean norm $\|c_z\|_2$ corresponds to the square-root of the minimum of cost function (20), i.e., $J_t = c_z^T c_z$. Note that for control, only the first elements corresponding to u_t or Δu_t are used from computed vector $\mathbb{U}_t$ (13) or (19), respectively.

In the previous paper of authors, [12], the point estimate corresponding to the centre of (12) was used in (28). The paper [13] extended these result both by using the incremental algorithm (24) and by considering the set estimate (12) without choice a particular point estimate. The transformation into (28) was performed successively for preselected points from the whole set. Subsequently, the realisation with the minimal value of $\|c_z\|_2$ was chosen as the result.

Here, we aim to optimize the choice of the relevant point estimate before optimization step to avoid the multiple optimisation run. For details on this choice see Sect. 4.3.

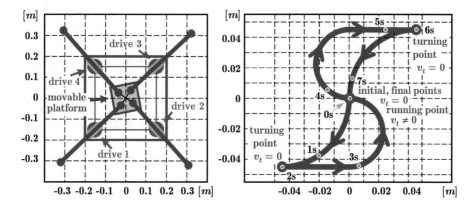

Fig. 1. Considered robot 'Moving Slide' and used testing trajectory [13].

Table 1. Overview of performed experiments.

Exp.	Control algorithm	State estimate	Fig. No	Mean E_t (31)	Max. E_t (31) (t)
1	Positional (23)	Central point (30)	Fig. 2	0.8977 mm	2.0958 mm (3.24 s)
2	Positional (23)	Set (12) [13]	Fig. 3, 4	0.9053 mm	2.3157 mm (4.70 s)
3	Positional (23)	Best point, Sect. 4.3	Fig. 5	0.9192 mm	2.2842 mm (3.24 s)
4	Incremental (24)	Central point (30)	Fig. 6	0.8248 mm	1.9193 mm (3.81 s)
5	Incremental (24)	Set (12) [13]	Fig. 7, 8	0.8317 mm	1.9069 mm (3.81 s)
6	Incremental (24)	Best point, Sect. 4.3	Fig. 9	0.8231 mm	1.8836 mm (3.81 s)

4 Experiments

4.1 Robot Model

To illustrate the proposed algorithm, we use the redundant planar parallel robot-manipulator [1]. It has a four-dimensional input u (four torques) and a three-dimensional output y (Cartesian positions of tool center point (TCP): x_{TCP} and y_{TCP}; and rotation angle φ_{TCP} of robot movable platform around the vertical axis), see Fig. 1. The dynamics of the robot can be described by a set of non-linear differential equations representing equations of motion. They are composed using Lagrange equations [3]

$$\ddot{y} = \mathsf{f}(\dot{y}, y) + \mathsf{g}(y)\, u \qquad (29)$$

where $y = [\ x_{TCP},\ y_{TCP},\ \varphi_{TCP}]^T$. The corresponding non-linear continuous-time state-space model can be transformed into the linear-like continuous-time state-space model by a special decomposition [24]. Then, using standard time discretisation and considering additive bounded disturbances, the LSU model (1), (2) is obtained [13] where the system state $x_t = [y_t^T, \dot{y}_t^T]^T$.

4.2 Experiment Setup

The controlled system is represented by the robot model (29) with an additive uniformly distributed output noise. The set state estimates (12) are obtained using the linearised robot model (1), (2) as described in Sect. 2. The noise bounds are set as follows: $\rho = 10^{-6}[m, m, rad, m\,s^{-1}, m\,s^{-1}, rad\,s^{-1}]^T$, $r = 10^{-3}[m, m, rad]^T$. The control parameters in (20) are set as follows: $N = 10$; $Q_{yw} = I$, $Q_u = 10^{-2}\,I$, $Q_{\Delta u} = 2.5 \cdot 10^{-2}\,I$, where I is the identity matrix of the appropriate order. The reference trajectory to be followed is depicted in Fig. 1.

The central point estimate, i.e. corresponding to the mean value of the set estimate (12), has the form [12]

$$\hat{x}_t = \frac{\underline{x}_t + \overline{x}_t}{2}. \tag{30}$$

The control error, i.e. a difference between the reference and a measured output, is defined as follows

$$E_t = \sqrt{\sum_{i=1}^{2}(y_i - w_i)^2} = \sqrt{\sum_{i=1}^{2} e_i^2}. \tag{31}$$

4.3 Search for the Best Point Estimate

Running the series of experiments based on the combination of boundary state values for corresponding couples: [position y_i + velocity $\dot{y}_i]_{i=1,2,3;}$, we have prove experimentally our initial hypothesis that the control results depend both on the choice of particular point from the set estimate (12) and on the previous state evolution that is related to the shape of the reference trajectory.

4.4 Results and Discussion

We have run series of experiments comparing the performance of the positional (23) and the incremental (24) output-feedback MPC algorithms with the following variants concerning the state point estimate: (i) a conservative choice (30) as presented in our previous paper [12], (ii) multiple optimisation using the scheme (28) with selection of points from (12) where the realization with a minimal loss c_z provides the required control input presented in our previous paper [13], (iii) offline setting of the "best" point state estimate as described in Sect. 4.3. A summary of the experiments is presented in Table 1.

In [13], the choice of the point estimate is performed in every optimisation step withing prediction horizon N. This strategy results in a discontinuous switching between subsequent point choices in control process, see the second halves of the control input courses in Fig. 3.

The proposed method of the "best" point state estimates applied to the incremental control algorithm (experiment 5) , delivers the least control error.

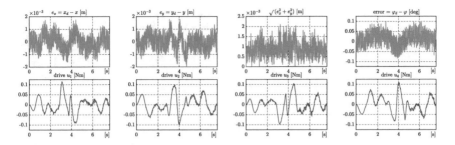

Fig. 2. Positional algorithm with mean value.

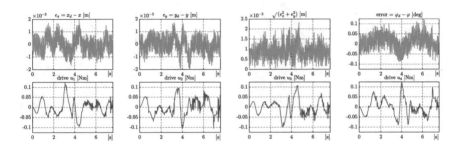

Fig. 3. Positional algorithm with selection according to loss value $c(z)$.

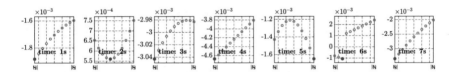

Fig. 4. Positional algorithm selected illustrative courses of loss values $c(z)$.

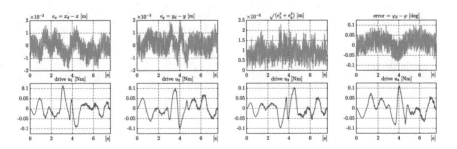

Fig. 5. Positional algorithm with selection based on prior information (1, 11, 6).

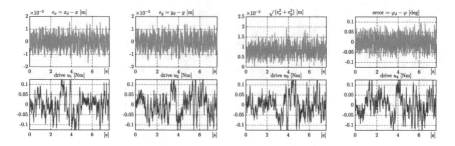

Fig. 6. Incremental algorithm with mean value.

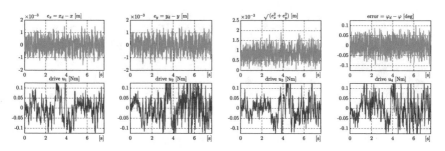

Fig. 7. Incremental algorithm with selection according to loss value $c(z)$.

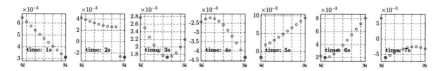

Fig. 8. Incremental algorithm selected illustrative courses of loss values $c(z)$.

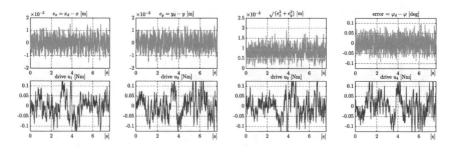

Fig. 9. Incremental algorithm with selection based on prior information (1, 11, 6).

5   Conclusion

The paper presents a method for the choice of the point state estimate used in the output-feedback MPC design. The results are compared with the previous authors results presented in [13] and [12]. In this paper, we proved experimentally our initial hypothesis that the control results depend both on the choice of particular point from the set estimate (12) and on the previous state evolution that is related to the shape of the reference trajectory.

Future work will focus on a theoretical justification of our hypothesis. We will aim to propose the optimal choice of point estimate based on available physical prior information about system motion behavior and on a physical substance of individual state variables e.g. position and corresponding velocity.

References

1. Belda, K.: Robotic device, Ct. 301781 CZ, Ind. Prop. Office (2010)
2. Belda, K.: On-line solution of system constraints in generalized predictive control design. In: 20th International Conference on Process Control, pp. 25–30 (2015)
3. Belda, K., Böhm, J., Píša, P.: Concepts of model-based control and trajectory planning for parallel robots. In: Klaus, S. (ed.) Proceedings of 13th IASTED International Conference on Robotics and Applications, pp. 15–20. Acta Press (2007)
4. Belda, K., Záda, V.: Predictive control for offset-free motion of industrial articulated robots. In: 2017 22nd International Conference Methods and Models in Automation and Robotics (MMAR), pp. 688–693 (2017)
5. Brunner, F.D., Müller, M.A., Allgöwer, F.: Enhancing output-feedback MPC with set-valued moving horizon estimation. IEEE Trans. Autom. Control 63(9), 2976–2986 (2018)
6. Combastel, C.: An extended zonotopic and gaussian Kalman filter (EZGKF) merging set-membership and stochastic paradigms: toward non-linear filtering and fault detection. Annu. Rev. Control. 42, 232–243 (2016)
7. Jirsa, L., Kuklišová Pavelková, L., Quinn, A.: Approximate Bayesian prediction using state space model with uniform noise. In: Gusikhin, O., Madani, K. (eds.) ICINCO 2018. LNEE, vol. 613, pp. 552–568. Springer, Cham (2020). https://doi.org/10.1007/978-3-030-31993-9_27
8. Jirsa, L., Kuklišová Pavelková, L., Quinn, A.: Approximate Bayesian prediction using state space model with uniform noise. In: Gusikhin, O., Madani, K. (eds.) ICINCO 2018. LNEE, vol. 613, pp. 552–568. Springer, Cham (2020). https://doi.org/10.1007/978-3-030-31993-9_27
9. Kárný, et al.: Optimized Bayesian Dynamic Advising: Theory and Algorithms. Springer, London (2005). https://doi.org/10.1007/1-84628-254-3
10. Khlebnikov, M.V., Polyak, B.T., Kuntsevich, V.M.: Optimization of linear systems subject to bounded exogenous disturbances: the invariant ellipsoid technique. Autom. Remote. Control. 72(11), 2227–2275 (2011)
11. Kögel, M., Findeisen, R.: Robust output feedback MPC for uncertain linear systems with reduced conservatism. IFAC-PapersOnLine 50(1), 10685–10690 (2017)
12. Kuklišová Pavelková, L., Belda, K.: Output-feedback model predictive control for systems under uniform disturbances. In: 2020 7th International Conference on Control, Decision and Information Technologies (CoDIT), pp. 897–902 (2020)
13. Kuklišová Pavelková, L., Belda, K.: Output-feedback MPC for robotic systems under bounded noise. In: Proceedings of the 18th International Conference on Informatics in Control, Automation and Robotics (ICINCO), pp. 574–582 (2021)
14. Lawson, C., Hanson, R.: Solving least squares problems. SIAM (1995)
15. Le, V., Stoica Maniu, C., Dumur, D., Alamo, T.: Robust tube-based constrained predictive control via zonotopic set-membership estimation. In: Proceedings of the IEEE Conference on Decision and Control, pp. 4580–4585 (12 2011)

16. Liu, Y., Zhao, Y., Wu, F.: Ellipsoidal state-bounding-based set-membership estimation for linear system with unknown-but-bounded disturbances. IET Control Theory Appl. **10**(4), 431–442 (2016)
17. Lorenzen, M., Cannon, M., Allgöwer, F.: Robust MPC with recursive model update. Automatica **103**, 461–471 (2019)
18. Luces, M., Mills, J.K., Benhabib, B.: A review of redundant parallel kinematic mechanisms. J. Int. Robot. Syst. **86**(2), 175–198 (2017)
19. Mammarella, M., Capello, E.: Tube-based robust MPC processor-in-the-loop validation for fixed-wing UAVs. J. Int. Robot. Syst. **100**(1), 239–258 (2020). https://doi.org/10.1007/s10846-020-01172-6
20. Meng, F., Liu, H., Shen, X., Wang, J.: Optimal prediction and update for box set-membership filter. IEEE Access **7**, 41636–41646 (2019)
21. Qiu, Q., Yang, F., Zhu, Y., Mousavinejad, E.: Output feedback model predictive control based on set-membership state estimation. IET Control Theory Appl. **14**(4), 558–567 (2020)
22. Smith, A.: Sequential Monte Carlo Methods in Practice. Springer, New York (2013). https://doi.org/10.1007/978-1-4757-3437-9
23. Tang, W., Wang, Z., Zhang, Q., Shen, Y.: Set-membership estimation for linear time-varying descriptor systems. Automatica **115**, 108867 (2020)
24. Valášek, M., Steinbauer, P.: Nonlinear control of multibody systems. In: Proceedings of Euromech, pp. 437–444 (1999)
25. Yadbantung, R., Bumroongsri, P.: Tube-based robust output feedback MPC for constrained LTV systems with applications in chemical processes. Eur. J. Control. **47**, 11–19 (2019)
26. You, F., Zhang, H., Wang, F.: A new set-membership estimation method based on zonotopes and ellipsoids. Trans. Inst. Meas. Control. **40**(7), 2091–2099 (2018)
27. Zenere, A., Zorzi, M.: Model predictive control meets robust Kalman filtering. IFAC-PapersOnLine **50**(1), 3774–3779 (2017)

Prediction of Overdispersed Count Data Using Real-Time Cluster-Based Discretization of Explanatory Variables

Evženie Uglickich[1($\boxtimes$)] and Ivan Nagy[1,2]

[1] Department of Signal Processing, Institute of Information Theory and Automation of the CAS, Pod vodárenskou věží 4, 18208 Prague, Czech Republic
{uglickich,nagy}@utia.cas.cz

[2] Faculty of Transportation Sciences, Czech Technical University, Na Florenci 25, 11000 Prague, Czech Republic
nagy@fd.cvut.cz

Abstract. The chapter focuses on the description of the relationship of the count variable and explanatory Gaussian variables. The cluster-based model is proposed, which is constructed on conditionally independent Gaussian clusters captured in real time using recursive algorithms of the Bayesian mixture estimation theory. The resulting model is expected to be used for predicting count data using real time Gaussian observations. The Poisson distribution of the count data is used as a basic model. However, in reality, count data often do not satisfy the Poisson assumption of equal mean and variance. For this case, five cluster-based Poisson-related models of overdispersed data have been studied. The experimental part of the chapter demonstrates a comparison of the prediction accuracy of the considered models with two theoretical counterparts for the case of weak and strong overdispersion with the help of simulations. The paper reports that the most accurate prediction in average has been provided by the cluster-based Generalized Poisson models.

Keywords: Cluster-based model · Count data · Overdispersion · Recursive Bayesian mixture estimation

1 Introduction

The chapter focuses on modeling and predicting count data variables generally described by the Poisson distribution. From a practical point of view, this task is required in application areas, dealing with random independent events observed with a constant intensity per time unit (e.g., social sciences, medicine, transportation, etc.) [1]. Specific examples of count variables considered per time unit include, e.g., a number of bankruptcies [2], aircraft shutdowns, specific diagnoses, server virus attacks [3], website users, customers [4,5], passengers [6], etc.

This work was supported by the project Arrowhead Tools, the project number ECSEL 826452 and MSMT 8A19009.

In application fields such as, e.g., transportation data analysis, the question of modeling count data depending on other observable variables arises. It can be met, for instance, in predicting passenger demand [6] or electric vehicle plugin intensity [7], etc. As the Poisson distribution does not assume general conditional form, the description of the relationship of the count Poisson-distributed variable with explanatory variables with the aim of constructing the data prediction model is a complicated task. For this task, the Poisson regression (as well as those using some of the Poisson-related distributions of the target variable) is one of the approaches most frequently met in literature e.g., [8–13]. In some sources, the application of linear regression techniques to Poisson-distributed count data due to high number of their possible realizations is also mentioned, see, for instance, [14].

Mixture models known as the universal approximation of nonlinear relation between variables [15] are used as well for the description of multimodal Poisson-distributed data. In this field, studies focusing on mixtures of Poisson distributions [16], mixtures of Poisson regressions [17, 18] and Poisson-gamma models [14] have been found.

In this chapter, the joint model of the count Poisson and multimodal Gaussian data is discussed. The similar problem is solved in the papers [19–22] via Gaussian-Poisson mixtures estimated with the help of the iterative expectation-maximization (EM) algorithm [23]. The prediction with the Poisson model is also considered in the papers [24, 25] and using the Bayesian methodology in [26]. The paper [27] proposed the approach, where the Poisson prediction probability function is constructed using the joint model of the target Poisson variable and Gaussian explanatory multivariate multimodal variable. The main features of the proposed algorithm [27] for the model estimation and data prediction are as follows: (i) the cluster-based discretization of Gaussian measurements, (ii) estimation of local Poisson models corresponding the discretization intervals, which include Gaussian data belonging to the detected clusters, and (iii) prediction of the target variable based on currently measured data discretized in real time. It should be noted that the cluster-based discretization was investigated, for example, in the papers [28–31]. In [27], it is based on the recursive mixture estimation [32, 33] under the Bayesian methodology.

The presented chapter is the extended version of the paper [27]. The aim of the chapter is to present the solution [27] for specific count data, which may be better described by special distributions based on the Poisson model. For example, to fit count observations with a high number of zeros, the zero-inflated Poisson model [34] as well as compound Poisson distributions [35] are used. Data without zeros can be fit by the zero-truncated Poisson distribution with a minimum at 1, see, e.g., [36]. Moreover, in reality, count data often do not satisfy the Poisson assumption of the equality of mean and variance, which means that the overdispersion or underdispersion of the data is observed.

This chapter focuses on overdispersed data as a more desired issue from the practical point of view; however, some of the used models are suitable for both the overdispersed and underdispersed count data [37]. The Generalized Poisson models [37, 38] and negative binomial regressions [39] are often used for the description of such data. In this chapter, the prediction of the count variable using the cluster-based discretization of continuous data [27] is considered for the mentioned Poisson-related distributions not requiring equidispersion, namely, the zero-truncated Poisson distribution (ZTP) and two Generalized Poisson models (GPM). The continuous Rayleigh distribution [40] has been used as well due to the shape of its probability density function similar to the Poisson one and high number of possible realizations of the count variable. The prediction with the Poisson, ZTP, GPM and Rayleigh distributions is experimentally compared with results obtained with the traditional Poisson and negative binomial regressions.

The layout of the chapter is organized as follows: Sect. 1.1 formulates the problem in general for the Poisson distribution. Section 2 provides the general solution with the cluster-based Poisson model. Its specification for overdispersed data described by the Poisson-related distributions can be found in Sect. 3. Section 4 discusses results of the experimental comparison of the mentioned models. Conclusions can be found in Sect. 5.

1.1   Problem Formulation

Let y be the count variable described by the Poisson distribution

$$f\left(y = y_t\right) = e^{-\lambda} \frac{\lambda^{y_t}}{y_t!} \qquad (1)$$

with the parameter λ, and let $y_t \in \{0, 1, \ldots\}$ be a realization of y observed on a multimodal system at discrete time instants $t = 1, 2, \ldots, T$.

Let $x = [x_1, x_2, \ldots, x_{N_x}]'$ be the multivariate Gaussian variable, and $x_t = [x_{1;t}, x_{2;t}, \ldots, x_{N_x;t}]'$ contains realizations of the vector x.

Observing the considered multimodal system, for the time instants $t > T$ the realizations x_t are still being measured, but y_t cannot be observed for the time $t > T$. Thus, the problem is verbally formulated as follows:

Predict the values of the dependent Poisson variable based on their relationship with realizations of explanatory Gaussian variables measured at real time $t > T$.

To solve the problem, the multimodality of realizations x_t is going to be used to describe the relationship of y and x under assumption of conditional independence of the individual variables $x_1, \ldots, x_{N_x}$ in the vector x. This will be done by the cluster-based discretization of the individual realizations from x_t and estimation of the local Poisson distributions using the data y_t measured at the same time instants as x_t in the detected clusters. The labels of the clusters will represent the discretized values of the Gaussian variables as it is shown in Fig. 1, which provides an illustrative scheme of the presented approach of the local model estimation. In this figure, for example, the blue-colored Poisson distribution covers count data, which have been measured at the same time with Gaussian data belonging to the cluster described by the blue-colored Gaussian distribution. The number 1 is the label of this blue cluster, which means that regarding this Gaussian distribution, the Poisson distribution numbered 1 exists.

This is believed to allow obtaining the conditional model of y depending on x in clusters in the form of the cluster-based Poisson model and use it for the prediction in real time. The solution is presented below.

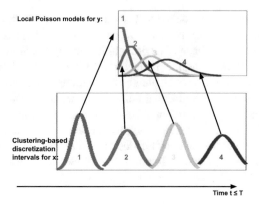

Fig. 1. The scheme of the estimation of the local Poisson distributions.

2   General Solution with Cluster-Based Poisson Model

Generally, the relationship between the dependent variable y and explanatory multi-variate variable $x = [x_1, \ldots, x_{N_x}]'$ is described by the joint distribution $f(y = y_t, x)$, where the denotation $f(y = y_t)$ relates to the discrete random variable y described by the Poisson distribution (1) with the realization y_t. For the continuous Gaussian multivariate variable x, the denotation of realizations is omitted. This joint distribution is decomposed with the help of the chain rule [41] in the following way:

$$f(y = y_t, x) = f(y = y_t | x_1, \ldots, x_{N_x}) f(x_1, \ldots, x_{N_x}). \tag{2}$$

The conditional distribution in the right side of (2) is the main focus of the work. Assuming the conditional independence of the individual variables $x_1, \ldots, x_{N_x}$, according to the Naive Bayes approach [49] it is derived as follows:

$$f(y = y_t | x_1, \ldots, x_{N_x}) \propto \frac{\prod_{l=1}^{N_x} f(y = y_t | x_l)}{f(y = y_t)^{N_x - 1}}. \tag{3}$$

Proof. According to the Bayes rule, see, e.g., [42], it holds

$$f(y = y_t | x_1, \ldots, x_{N_x}) = \frac{f(x_1, \ldots, x_{N_x} | y = y_t) f(y = y_t)}{f(x_1, \ldots, x_{N_x})}. \tag{4}$$

Assuming the conditional independence of the individual $x_1, \ldots, x_{N_x}$, the right side of the above relation results in

$$\frac{\prod_{l=1}^{N_x} f(x_l | y = y_t) f(y = y_t)}{f(x_1, \ldots, x_{N_x})}. \tag{5}$$

Next, applying the Bayes rule to the individual conditional distributions in (5) gives

$$f(x_l | y = y_t) = \frac{f(y = y_t | x_l) f(x_l)}{f(y = y_t)}, \quad \forall l \in \{1, \ldots, N_x\}. \tag{6}$$

Substituting (6) into (5) provides

$$
\prod_{l=1}^{N_x} \left[\frac{f(y = y_t|x_l)f(x_l)}{f(y = y_t)} \right] \frac{f(y = y_t)}{f(x_1, \dots, x_{N_x})}
$$
$$
= \frac{\prod_{l=1}^{N_x} f(y = y_t|x_l) \prod_{l=1}^{N_x} f(x_l)}{f(y = y_t)^{N_x}} \frac{f(y = y_t)}{f(x_1, \dots, x_{N_x})} \tag{7}
$$
$$
= \frac{\prod_{l=1}^{N_x} f(y = y_t|x_l)}{f(y = y_t)^{N_x-1}} \frac{\prod_{l=1}^{N_x} f(x_l)}{f(x_1, \dots, x_{N_x})} \propto \frac{\prod_{l=1}^{N_x} f(y = y_t|x_l)}{f(y = y_t)^{N_x-1}},
$$

where in view of the modeled variable y in the left side of (3), $\frac{\prod_{l=1}^{N_x} f(x_l)}{f(x_1, \dots, x_{N_x})}$ is a constant value with the substituted realizations of each x_l.

In the numerator of the relation (3), the scalar models $f(y = y_t|x_l)$, $l \in \{1, \dots, N_x\}$ express the dependence of y on the individual variables x_l. Since (i) x_l are continuous and (ii) the Poisson distribution does not have a general conditional form, a discrete-valued condition brings easier solution in the form of the Poisson distribution defined for each discrete value in the condition. That is why it is proposed to approximate each of the models $f(y = y_t|x_l)$ in (3) by the Poisson distribution existing for each discretized value of each variable x_l as follows:

$$
f(y = y_t|x_l) \approx f(y = y_t|\tilde{x}_l = i) \equiv f_i(y = y_t), \tag{8}
$$

where $\tilde{x}_l$ is the discretized variable of x_l and it has realizations $i \in \{1, 2, \dots, N_{c;l}\}$. The recursive cluster-based discretization is used in the paper, which means that the value i of $\tilde{x}_l$ is a label of the cluster, to which the observations of the corresponding x_l belong, and $N_{c;l}$ is the number of its clusters.

In this way, to discretize the data here means to find the clusters. For this aim, the scalar marginal distributions $f(x_l)$ are used. Each of them is approximated by the mixture of $N_{c;l}$ Gaussian components $\mathcal{N}_i(x_{l;t}; \theta_{i;l}, r_{i;l})$ with the collection of the unknown expectations $\theta_{i;l}$. Their variances $r_{i;l}$ are set known and fixed in order to locate tops of the data hills. The cluster-based discretization is presented below.

2.1 Cluster-Based Discretization of Explanatory Variables

The realizations $x_t = [x_{1;t}, \ x_{2;t}, \ \dots \ x_{N_x;t}]'$ observed at time $t = 1, \dots, T$ are used for this part of the proposed approach. Here, the aim is to discretize these realizations to the clusters and use labels of the clusters as the values of $\tilde{x}_l$ for (8). Specifically, this task covers the estimation of individual expectations $\theta_{i;l}$ identifying each i-th Gaussian component of each l-th variable and their labels $\tilde{x}_l$.

The cluster-based discretization of realizations of each x_l is based on the recursive Bayesian mixture estimation methodology [32,33,43]. Its significant advantages are: (i) the recursive running of the clustering algorithms based on actually measured data, which is suitable in view of the prediction task formulated in Sect. 1.1, and (ii) the simple and efficient mixture initialization of the univariate components based on prior knowledge. For the discretization, it worth setting relatively bigger number of components, for instance, 10 or 15, in order not to lose the important information in the data.

According to [32,43], the joint distribution of unknown expectations $\theta_{i;l}$ and discretized values $\tilde{x}_l$ is constructed for the i-th Gaussian component of the l-th variable using the Bayes and chain rules as follows: $\forall i \in \{1, 2, \ldots, N_{c;l}\}, \forall l \in \{1, \ldots, N_x\}$

$$
\begin{aligned}
f(\tilde{x}_{l;t} = i, \theta_{i;l} | x_l(t)) &\propto f(x_{l;t}, \tilde{x}_{l;t} = i, \theta_{i;l} | x_l(t-1)) \\
&= f(x_{l;t} | \tilde{x}_{l;t} = i, \theta_{i;l}, x_l(t-1)) f(\tilde{x}_{l;t} = i | \theta_{i;l}, x_l(t-1)) f(\theta_{i;l} | x_l(t-1)) \\
&= \underbrace{f_i(x_{l;t} | \theta_{i;l}, x_l(t-1))}_{\mathcal{N}_i(x_{l;t};\theta_{i;l},r_{i;l})} \underbrace{f_i(\theta_{i;l} | x_l(t-1))}_{\mathcal{N}_i(\theta_{i;l})} \underbrace{f_i(\tilde{x}_{l;t} = i | x_l(t-1))}_{\mathcal{U}_i(\tilde{x}_l)} \\
&\propto \underbrace{f_i(x_{l;t} | \theta_{i;l}, x_l(t-1))}_{\mathcal{N}_i(x_{l;t};\theta_{i;l},r_{i;l})} \underbrace{f_i(\theta_{i;l} | x_l(t-1))}_{\mathcal{N}_i(\theta_{i;l})}
\end{aligned}
\tag{9}
$$

where

- $\tilde{x}_{l;t} = i$ denotes the label of the i-th cluster described by the i-th component and detected at time t;
- $x_l(t) = \{x_{l;0}, x_{l;1}, \ldots, x_{l;t}\}$ is a collection of all data of x_l up to the time t with the prior knowledge $x_{l;0}$;
- $\mathcal{N}_i(x_{l;t}; \theta_{i;l}, r_{i;l})$ is the i-th Gaussian component of x_l;
- $\mathcal{N}_i(\theta_{i;l})$ is the prior Gaussian probability density function (pdf);
- $\mathcal{U}_i(\tilde{x}_l)$ is the prior uniform distribution of the discretized variable $\tilde{x}_l$ of x_l, which is a constant in view of the left side of the expression;
- and $\tilde{x}_l$ and $\theta_{i;l}$ are assumed to be mutually independent.

To obtain the estimate of the actual component (i.e., to derive the posterior distribution of $\tilde{x}_l$ based on the current data), the decomposed joint distribution from the right side of (9) is marginalized over the expectations $\theta_{i;l}$

$$
f_i(\tilde{x}_{l;t} = i | x_l(t)) \propto \int_{\theta^*} \underbrace{f_i(x_{l;t} | \theta_{i;l}, x_l(t-1))}_{\mathcal{N}_i(x_{l;t};\theta_{i;l},r_{i;l})} \underbrace{f_i(\theta_{i;l} | x_l(t-1))}_{\mathcal{N}_i(\theta_{i;l})} \, d\theta_{i;l},
\tag{10}
$$

where θ^* means the entire definition space of the expectation. Here, the approximation with the help of the Dirac delta function $\delta(\theta_{i;l}, \hat{\theta}_{i;l;t-1})$, which allows to substitute the prior point estimates $\hat{\theta}_{i;l;t-1}$ of the expectations into the Gaussian components, provides a significant simplification of the solution, see, e.g., [33]. Due to this approximation, the integral in the right side of (10) can be denoted by

$$
q_{i;l} = \int_{\theta^*} \underbrace{f_i(x_{l;t} | \theta_{i;l}, x_l(t-1))}_{\mathcal{N}_i(x_{l;t};\theta_{i;l},r_{i;l})} \underbrace{f_i(\theta_{i;l} | x_l(t-1))}_{\mathcal{N}_i(\theta_{i;l})} \, d\theta_{i;l} \approx \underbrace{f_i(x_{l;t} | \hat{\theta}_{i;l;t-1}, x_l(t-1))}_{\mathcal{N}_i(x_{l;t};\hat{\theta}_{i;l;t-1},r_{i;l})},
\tag{11}
$$

which is the value of the i-th Gaussian component pdf with the substituted prior point estimate $\hat{\theta}_{i;l;t-1}$ and current realization $x_{l;t}$. After the normalization

$$
m_{i;l} = \frac{q_{i;l}}{\sum_{k=1}^{N_{c;l}} q_{k;l}},
\tag{12}
$$

it provides the normalized proximity [33] of the current realization $x_{l;t}$ to the i-th component of the l-th variable, $\forall i \in \{1, 2, \ldots, N_{c;l}\}$, $\forall l \in \{1, \ldots, N_x\}$. Then the searched posterior distribution (10) of the discretized variable $\tilde{x}_l$ is

$$f_i(\tilde{x}_{l;t} = i|x_l(t)) \propto m_{i;l}, \tag{13}$$

which represents the probability of the membership of the current realization $x_{l;t}$ to each of the $N_{c;l}$ components of the variable x_l at time t. The point estimate of the discretized variable $\tilde{x}_l$ at time t, which labels the Poisson distributions is the argument of the maxima of (13)

$$\tilde{x}_{l;t} = \arg\max_i f_i(\tilde{x}_{l;t} = i|x_l(t)). \tag{14}$$

In this way, the continuous realizations of each variable x_l at time t are now represented by the discrete labels $\tilde{x}_l$ of its clusters described by the components.

The cluster-based discretization (10)–(14) is performed in the recursive way similarly to [32,43] to re-compute the point estimates of the expectations $\theta_{i;l}$ and discretized variables $\tilde{x}_l$ at time t. For this end, statistics of the prior Gaussian pdfs $\{V_{i;l;t-1}, \kappa_{i;l;t-1}\}$ conjugate to the i-th individual components describing the clusters of the variables x_l are updated with the new realizations weighted by the proximities as follows:

$$V_{i;l;t} = V_{i;l;t-1} + m_{i;l}x_{l;t}, \quad \kappa_{i;l;t} = \kappa_{i;l;t-1} + m_{i;l}. \tag{15}$$

They are used for re-computing the point estimates of the expectations at time t

$$\hat{\theta}_{i;l;t} = \frac{V_{i;l;t}}{\kappa_{i;l;t}}, \quad \forall i \in \{1, 2, \ldots, N_{c;l}\}, \forall l \in \{1, \ldots, N_x\} \tag{16}$$

to be used in (11). The recursive discretization (10)–(14) runs until the time $t = T$, i.e., for a data set with realizations of x_l and y observed at the same time. As a result, the point estimates of $\tilde{x}_l$ (14), which discretize the Gaussian realizations, are now used to label the Poisson distributions (8). This is explained in the next section.

2.2 Estimation of the Local Poisson Distributions on Clusters

The cluster-based Poisson model is constructed using observations of the individual variables x_l discretized to their clusters and the realizations y_t measured at the same time instants as the clustered data up to the time $t = T$. The searched Poisson distributions (8) labeled by the point estimates i of the discretized variable $\tilde{x}_l$ from (14)

$$\underbrace{f_i(y = y_t)}_{\mathcal{P}oi(y_t; \lambda_{i;l})} \tag{17}$$

are given by the point estimates of their parameters $\lambda_{i;l}$, which are the averages of the realizations y_t measured simultaneously with the data $x_{l;t}$ belonging to each i-th cluster, $\forall i \in \{1, 2, \ldots, N_{c;l}\}, \forall l \in \{1, \ldots, N_x\}$.

2.3 Prediction with the Cluster-Based Poisson Model

According to the problem formulation in Sect. 1.1, the prediction of the realizations y_t for time $t > T$ should be based on the new data $x_{l;t}$. The cluster-based Poisson model is used for the prediction in real time $t > T$ as follows:

1. The current realizations of each x_l are measured;
2. The actual proximities $m_{i;l}$ to their components are computed;
3. The Poisson distributions (17) estimated in Sect. 2.2 are used to obtain their weighted averages $\forall l \in \{1, \ldots, N_x\}$

$$f(y = y_t | x_l) = \sum_{i=1}^{N_{c;l}} m_{i;l} \underbrace{f_i(y = y_t)}_{\mathcal{P}oi(y_t; \lambda_{i;l})}, \tag{18}$$

which is the resulting cluster-based Poisson model for the variable x_l.
4. To obtain the predictive model, the individual models (18) with each x_l in the condition are substituted into (3), where the denominator contains values of the Poisson distribution for all previously available realizations y_t.
5. The result of (3) is either the predictive distribution $f(y = y_t | x_1, \ldots, x_{N_x})$ or the point prediction of the realization y_t in real time $t > T$ as the argument of the maxima

$$\hat{y}_t = \arg\max_j f(y = y_t | x_1, \ldots, x_{N_x}), \quad j \in \{0, 1, \ldots, N_y\}, \tag{19}$$

where N_y is the maximum observed value of the count variable. This is the main result according to the problem formulation in Sect. 1.1.

3   Cluster-Based Models of Overdispersed Count Data

This section is devoted to the specification of the solution presented in Sect. 2 to distributions describing count observations, where the Poisson assumption of the equality of mean and variance is violated, i.e., overdispersion or underdispersion is present. In real applications, the overdispersion is frequently met, that is why this section focuses on this case. The distributions considered in the section are suitable for modeling overdispersed count data because, even though they are based on the Poisson distribution, they allow the variance to be a function of mean as a rule through an additional dispersion parameter.

In this chapter, the following distributions are used to describe the count variable y defined in Sect. 1.1, which is assumed to be overdispersed.

3.1   Cluster-Based Zero-Truncated Poisson Model

The zero-truncated Poisson distribution (ZTP) is taken in the chapter using the following denotation

$$f(y = y_t | y_t > 0) = \frac{\lambda^{y_t}}{(e^\lambda - 1) y_t!} \tag{20}$$

with the parameter λ and the mean

$$E[y] = \frac{\lambda}{1 - e^{-\lambda}} \tag{21}$$

and the variance as a function of the mean

$$D[y] = \frac{\lambda + \lambda^2}{1 - e^{-\lambda}} - \frac{\lambda^2}{(1 - e^{-\lambda})^2} = E[y](1 + \lambda - E[y]). \tag{22}$$

The ZTP distribution (20) excludes zero realizations $y_t = 0$ of the variable y. For this reason, the data recoding via their shifting to a minimum at 1 is necessary to use this model instead of the Poisson distribution (1) in the presented approach. For the prediction accuracy evaluation, the re-shift at zero must be applied.

The ZTP model (20) is estimated locally on the clusters instead of the Poisson distributions in (17) (see Sect. 2.2) and substituted into (18) in the prediction part of the algorithm (see Sect. 2.3). According to [44], the maximum likelihood estimation of the parameter λ is obtained numerically via solving the equation (21) using the sample mean instead of $E[y]$. This model has been tested for the approach in Sect. 4; however, for bigger values of the parameter λ, the mean is approximately equal to this parameter, and consequently, $D[y] \approx E[y]$ in (22) is obtained. The proposed approach with the ZTP model (20) is therefore restricted for small values of λ.

3.2 Cluster-Based Consul's Generalized Poisson Model

The Consul's Generalized Poisson model (GP1) [38] is used in this work in the form of the following probability function

$$f(y = y_t) = \frac{e^{-(\lambda + a y_t)}(\lambda + a y_t)^{y_t - 1}}{y_t!}, \tag{23}$$

where λ is the parameter of GP1 and a is the dispersion parameter. The mean and variance of the distribution are respectively

$$E[y] = \frac{\lambda}{1 - a}, \quad D[y] = \frac{\lambda}{(1 - a)^3}. \tag{24}$$

The dispersion parameter a is estimated according to the following formula [38,45,46]

$$a = \frac{\sum_{t=T+1}^{\tilde{T}} \left(\frac{|y_t - \hat{y}_t|}{\sqrt{\hat{y}_t}} - 1 \right)}{(\tilde{T} - T) - N_x - 1}, \tag{25}$$

where $\hat{y}_t$ are the predictions obtained either with the cluster-based Poisson model or the traditional Poisson regression for the testing data set of y_t and x_t measured for time $t = T + 1, \ldots, \tilde{T}$, and N_x is the number of the explanatory variables. Further, the point estimate of a is substituted along with the sample mean instead of $E[y]$ in (24) to obtain the estimate of the parameter λ.

Similarly, within the presented approach, the parameter λ of the GP1 model is estimated locally on the clusters instead of the Poisson distributions in (17) from Sect. 2.2 and substituted into (18) in the prediction part of the algorithm (see Sect. 2.3).

3.3   Cluster-Based Famoye's Generalized Poisson Model

Famoye's Generalized Poisson model (GP2) [38] is the next distribution tested within the bounds of the presented approach. It has the form

$$f(y = y_t) = \frac{\lambda}{1 + a\lambda} \frac{(\lambda + ay_t)^{y_t-1}}{y_t!} e^{\frac{-\lambda(1+ay_t)}{1+a\lambda}}, \tag{26}$$

where λ is the parameter of the GP2 model. This distribution has the following mean and variance:

$$E[y] = \lambda, \quad D[y] = \lambda(1 + a\lambda)^2, \tag{27}$$

while the dispersion parameter a is estimated according to [38,45,46] as follows:

$$a = \frac{\sum_{t=T+1}^{\tilde{T}} \left(\frac{|y_t - \hat{y}_t|}{\sqrt{\hat{y}_t}} - 1 \right) \frac{1}{\hat{y}_t}}{(\tilde{T} - T) - N_x - 1}, \tag{28}$$

with $\hat{y}_t$ and the rest of denotations defined similarly as for (25). Here, the estimation of the parameter λ is straightforward due to the use of the sample mean in (27) locally on the clusters similarly to (17). In the prediction part, it is again substituted into (18) according to Sect. 2.3.

3.4   Cluster-Based Rayleigh Model

The continuous Rayleigh distribution suitable for non-negative data has the shape of the probability density function relatively close to the Poisson distribution. With a high number of possible realizations of the count data it can serve as an approximation of the modeled variable. That is why it is used in this work to test the proposed approach. The Rayleigh distribution has the following form

$$f(y = y_t) = \frac{y_t}{\sigma^2} e^{-y_t^2/(2\sigma^2)}, \tag{29}$$

where σ is the parameter of the Rayleigh distribution. Its mean and variance are approximately connected with σ through the following relations:

$$E[y] \approx 1.253\sigma, \quad D[y] \approx 0.429\sigma^2. \tag{30}$$

The parameter estimation is performed according to [47] using

$$\sigma \approx \sqrt{\frac{1}{2T} \sum_{t=1}^{T} y_t^2}, \tag{31}$$

which is calculated locally on the clusters similarly to (17) and then substituted into (18) instead of the Poisson distribution according to Sect. 2.3.

3.5   Theoretical Counterparts

The distributions introduced in the above sections are used as the cluster-based models instead of the Poisson distribution according to Sect. 2. Results of predicting based on these models are compared with (i) the Poisson regression (which can be used if the equality of mean and variance is not violated) and (ii) the negative binomial regression suitable for overdispersed data.

Poisson Regression. The Poisson regression assumes that the relation between the Poisson parameter $\lambda = E[y] = D[y]$ and realizations of the explanatory variables in the vector x_t has the following form

$$\ln(\lambda) = \theta' x_t = b_0 + b_1 x_{1;t} + b_2 x_{2;t} + \ldots + b_{N_x} x_{N_x;t}, \tag{32}$$

where the vector $\theta = [b_0 \; b_1 \; b_2 \; \ldots \; b_{N_x}]'$ contains regression coefficients. Here, they are estimated with the help of the linearization of the Poisson regression and subsequent application of the least square estimator in the following form:

$$\underbrace{\begin{bmatrix} \ln(y_1) \\ \ln(y_2) \\ \ldots \\ \ln(y_T) \end{bmatrix}}_{Y} = \underbrace{\begin{bmatrix} 1 & x_{1;1} & \ldots & x_{N_x;1} \\ 1 & x_{1;2} & \ldots & x_{N_x;2} \\ \ldots & \ldots & \ldots & \ldots \\ 1 & x_{1;T} & \ldots & x_{N_x;T} \end{bmatrix}}_{X} \underbrace{\begin{bmatrix} b_0 \\ b_1 \\ \ldots \\ b_{N_x} \end{bmatrix}}_{\theta}. \tag{33}$$

The vector of the regression coefficients is estimated using

$$\theta = (X'X)^{-1} X'Y, \tag{34}$$

see, e.g., [41]. The prediction is then obtained using the observations of the explanatory variables for $t > T$ as follows:

$$\hat{y}_t = e^{\theta' x_t}. \tag{35}$$

Negative Binomial Regression. The negative binomial regression is a generalization of the Poisson regression, which does not require the assumption of equidispersion to be satisfied. It assumes that the count variable y follows the negative binomial distribution (NB)

$$f(y = y_t) = \binom{y_t + r - 1}{r - 1} (1 - p)^{y_t} p^r, \tag{36}$$

where p and r are parameters of the distribution. The NB distribution can be also defined with the alternative parametrization through the mean μ and variance σ^2

$$f(y = y_t) = \binom{y_t + \frac{\mu^2}{\sigma^2 - \mu} - 1}{y_t} \left(\frac{\sigma^2 - \mu}{\sigma^2}\right)^{y_t} \left(\frac{\mu}{\sigma^2}\right)^{\mu^2/(\sigma^2 - \mu)}. \tag{37}$$

The NB regression with the target variable y is given by the relation

$$\mu = \exp\{\ln t + b_0 + b_1 x_{1;t} + b_2 x_{2;t} + \ldots + b_{N_x} x_{N_x;t}\}, \tag{38}$$

or

$$\ln(\mu) = \theta' \begin{bmatrix} t \\ x_t \end{bmatrix} = t + b_0 + b_1 x_{1;t} + b_2 x_{2;t} + \ldots + b_{N_x} x_{N_x;t}, \tag{39}$$

where, similarly to the Poisson regression, the vector $\theta = [1 \ \ b_0 \ \ b_1 \ \ b_2 \ \ \ldots \ \ b_{N_x}]'$ contains regression coefficients. After the linearization, the regression coefficients are estimated similarly to (33–34) with the corresponding arrangements. The prediction is done according to (35) using the extended vector of the explanatory variables instead of x_t.

Two types of the NB regression is distinguished [48]: the first one denoted by NB1 has the variance

$$D[y] = E[y](1 + a_{nb1}), \tag{40}$$

and the second one denoted by NB2 has the variance

$$D[y] = E[y] + a_{nb2} * E[y]^2, \tag{41}$$

while the mean is the same for both of them. According to [48], the dispersion parameter a_{nb1} for the NB1 model with the variance (40) is estimated using auxiliary least squares (similarly to (34)) with the equation

$$\frac{(y_t - \hat{y}_t)^2 - \hat{y}_t}{\hat{y}_t} = a_{nb1} + 0, \tag{42}$$

where $\hat{y}_t$ is the prediction obtained with the Poisson regression in (35). The column of the values of the left side of this equation for each time instant t is denoted by Y; the unit vector of the corresponding dimension is denoted by X and the intercept is equal to zero.

The dispersion parameter a_{nb2} for the NB2 regression with the variance (41) is estimated in a similar way, see, e.g., [48], with the help of auxiliary least squares solved for the equation

$$\frac{(y_t - \hat{y}_t)^2 - \hat{y}_t}{\hat{y}_t} = a_{nb2}\hat{y}_t + 0, \tag{43}$$

where the columns of the observations and predictions are denoted as follows: $Y = \frac{(y_t - \hat{y}_t)^2 - \hat{y}_t}{\hat{y}_t}$, $B_1 = a_{nb2}$, $X = \hat{y}_t$, $B_0 = 0$, and it holds $Y = B_1 x + B_0$.

In this work, the NB1 regression with the variance (40) is used.

Remark: The cluster-based NB model has been also tested with the bounds of the presented approach, assuming the usage of the NB distribution instead of the Poisson model according to Sect. 2. However, due to the computational complexity caused by complex numbers because of the negative number in factorial this attempt has been stopped so far.

The following section demonstrates results of experiments with the introduced models.

4   Experiments

The presented experiments have been conducted in a free and open source programming environment Scilab (www.scilab.org) aimed at engineering and scientific computations. The aim of the experiments was to verify the presented algorithm with the simulated data and compare the accuracy of predicting the count variable y using all of the given models.

4.1   Simulations

3000 data items have been simulated within an individual data set. Each of the data sets contains realizations of the count variable y and four-dimensional multimodal Gaussian variable $x = [x_1, x_2, x_3, x_4]'$, i.e., $N_x = 4$. For the cluster-based discretization of the Gaussian single variables, a different number of components have been chosen during the simulation for each of them. Four variables had $10, 12, 3$ and 15 components respectively in order to test both the high and low number of components.

Realizations of the count variable y have been simulated so that to have different values of the Poisson parameter λ for all of the clusters of each Gaussian variable. The equality of the mean and variance of the resulting count simulations were not supposed due to their multimodal nature, otherwise a single Poisson distribution would be enough to describe the data, which is irrelevant within the bounds of the considered problem. However, the interesting issue is the variances of the overdispersed count data used for the estimation of parameters of the considered distributions corresponding to the clusters of the Gaussian variables.

Approximately 100 data sets have been simulated for the case of weaker and stronger overdispersion of count data. Examples of the data mean and variances of both the cases are given in Table 1, while examples of histograms are shown in Fig. 2.

Table 1. Examples of the data means and variances of the data sets.

Data overdispersion	Mean	Variance
Weak	4.9023333	6.8630822
Strong	34.140667	275.99821

The following section demonstrates results of the prediction of the count variable according to the presented algorithm from Sect. 2 and its comparison with the Poisson and NB regressions.

4.2   Results and Discussion

Prediction of Weakly Overdispersed Count Data. In this section, the presented algorithm has been applied to approximately 100 data sets of 3000 values of x and y, where the values of y have been generated with the weak overdispersion. For each data set, 2800 data items have been utilized for the cluster-based discretization and local model estimation according to Sects. 2.1 and 2.2, i.e., $T = 2800$. The rest of 200 values have been used for the prediction algorithm according to Sect. 2.3 with the given models.

Weakly overdispersed data

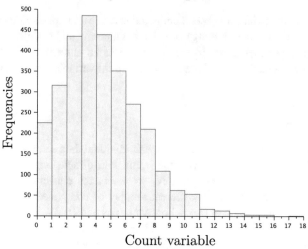

Strongly overdispersed data

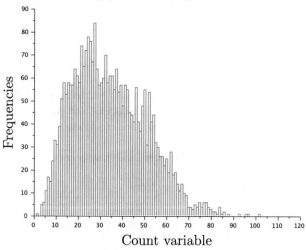

Fig. 2. Histogram examples of data with a weak (top) and strong (bottom) overdispersion.

The results of the experiments have been compared from the prediction accuracy point of view with the help of the following criteria:

$$\text{root-mean-square error RMSE} = \sqrt{\frac{\sum_{t=T+1}^{\tilde{T}}(y_t - \hat{y}_t)^2}{\tilde{T} - T}}, \tag{44}$$

where $\hat{y}_t$ stands for the predicted value obtained with each model, and $\tilde{T} = 3000$;

$$\text{relative-prediction error RPE} = \frac{D[y_t - \hat{y}_t]}{D[y_t]}, \qquad (45)$$

where D denotes variance, and $t = T + 1, \ldots, \tilde{T}$;

$$\text{Akaike information criterion AIC} = -2\ln(\mathcal{L}) + 2N_x \qquad (46)$$

and

$$\text{Bayesian information criterion BIC} = -2\ln(\mathcal{L}) + N_x \ln(\tilde{T} - T), \qquad (47)$$

where $\mathcal{L}$ denotes the likelihood.

Table 2 demonstrates the comparison of the prediction accuracy with the weakly overdispersed data among all of the considered models, including the cluster-based Poisson model and Poisson regression. Due to the smaller values of the count variable, the cluster-based ZTP model has been used as well. The average values of RMSE, RPE, AIC and BIC calculated on the used sets of simulations are presented.

Table 2. The average prediction accuracy for weakly overdispersed data.

Model	RMSE	RPE	AIC	BIC
Cluster-based Poisson model	2.7724659	0.6870497	974.28897	987.50218
Cluster-based ZTP model	2.7760525	0.6887404	1001.3196	1014.5328
Cluster-based GP1 model	2.7965864	0.6901744	1802.6274	1815.8406
Cluster-based GP2 model	2.9883688	0.701766	1082.8114	1096.0246
Cluster-based Rayleigh model	5.5489679	0.7176861	1259.2657	1272.4789
Poisson regression	3.2819422	0.8568735	1131.8207	1148.3372
NB1 regression	3.1889111	0.8374878	1067.8391	1084.3556

It can be seen in Table 2 that all of the compared models provide the relatively high prediction error in view of the small mean and variance of the weakly dispersed data with the average range 20 and the minimum at 0. However, to evaluate the comparison among the obtained results, it can be noted that regarding RMSE and RPE, the cluster-based models (except the Rayleigh model) show the higher prediction accuracy than the traditional Poisson and NB1 regressions. The lowest RMSE, RPE, AIC and BIC have been obtained naturally with the cluster-based Poisson model. It is explained by the weak overdispersion, which means that the data distributions obtained on clusters were close to the Poisson one.

If one omits the cluster-based Poisson model and Poisson regression in view of the overdispersion, the cluster-based ZTP model has the lowest RMSE, RPE, AIC as well as BIC and shows improvements against the NB1 regression. However, it is sensitive to bigger values of the count data (see Sect. 3.1).

From this point of view, the cluster-based GP2 model shows the most balanced improvements of the prediction accuracy in the comparison with the rest of the models for the case of the weakly overdispersed data: it has the lowest RMSE and RPE. Its AIC and BIC are insignificantly higher than the NB1 regression results.

The comparison of variances of the count data captured on clusters of individual Gaussian variables during the discretization shows that they are close to the variances of the GP2 distributions on these clusters. However, the corresponding test of hypothesis has not been performed because of the small number of components. For the illustration, Table 3 demonstrates the means, variances of data and GP2 variances on clusters of the variables x_1, x_2 and x_3 obtained with one of the data sets. The variable x_4 was omitted to save space. The histograms of the count data collected on the clusters of the variable x_3 are shown in Fig. 3 as an example of the distributions on the clusters.

Table 3. The mean, variance of data and the GP2 variances on clusters of the variables.

Mean	Variance of data	GP2 variance on clusters of x_1
3.2553606	3.4327105	3.1201919
3.7605893	4.5366727	3.5805055
3.9872204	4.9470364	3.7849268
4.3664596	5.196118	4.1241545
4.6079734	5.7258029	4.3383356
4.45	5.9447368	4.1984032
5.0644068	6.9176064	4.7391929
4.2	3.2	3.9756952
5.2034483	7.1107147	4.8602884
5.8913043	7.4323671	5.4524133
Mean	Variance of data	GP2 variance on clusters of x_2
2.952	3.3686452	2.8407398
2.7964602	2.9492731	2.6965651
3.35	3.2589744	3.2069019
3.7756654	4.6097872	3.5941437
3.8919861	5.0407397	3.6991806
4.1469194	5.1565107	3.9282105
4.2212121	5.285263	3.9946516
4.2061856	5.9054538	3.9812241
4.4304348	6.019157	4.1810295
4.5490909	6.3725813	4.2862496
4.689243	5.6870438	4.4100846
4.9888889	6.6206939	4.6732235
Mean	Variance of data	GP2 variance on clusters of x_3
3.5916667	4.4086896	3.4273074
4.3343109	5.7217864	4.0955355
4.6511628	6.2274282	4.3764856

Prediction of Strongly Overdispersed Count Data. In this section, results of predicting of strongly overdispersed count data are discussed. In this part of the experiments, the cluster-based Poisson and ZTP models as well as Poisson regression were not used due to the violated Poisson assumption and higher values of the counts (see Sect. 3.1). The experiments have been conducted under the same conditions as specified in the previous section.

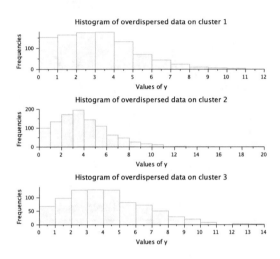

Fig. 3. The histograms of the count data collected on the clusters of the variable x_3.

Table 4. The average prediction accuracy for strongly overdispersed data.

Model	RMSE	RPE	AIC	BIC
Cluster-based GP1 model	5.3527321	0.2450608	2586.4615	2599.6747
Cluster-based GP2 model	10.156487	0.4225181	2988.1786	3001.3918
Cluster-based Rayleigh model	8.5742296	0.5272263	6235.3781	6248.5913
NB1 regression	8.6107072	0.6270219	1784.7504	1801.267

Table 4 provides the comparison of the average values of RMSE, RPE, AIC and BIC of the models calculated on the sets of simulations. Here, it can be seen, that the lowest RMSE and RPE have been obtained with the cluster-based GP1 model. However, its AIC and BIC are in the second place after the NB1 regression.

For the illustration, a fragment of the prediction is shown in Fig. 4, where the GP1 predictions follow the simulations. Figure 5 compares the histograms of the data from one of the testing sets with the GP1 and NB1 predictions. Both the simulations and GP1 predictions have the values approximately from 5 to 50, while the NB1 histogram provides the values from 10 to 44.

For the strongly overdispersed data, there is not the similar concordance in the variances of data and distributions on clusters as observed for weak overdispersion. It is explained by a high measure of the uncertainty in such data.

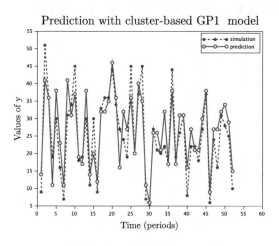

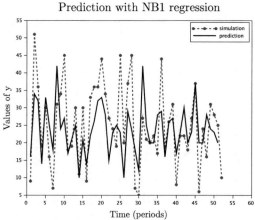

Fig. 4. The prediction with the cluster-based GP1 (top) and NB1 (bottom) models.

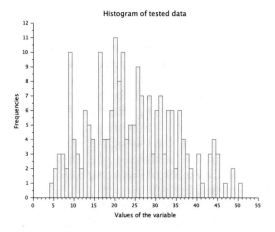

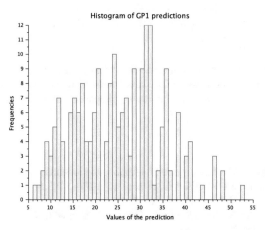

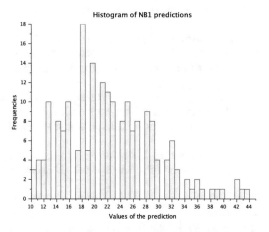

Fig. 5. The histograms of data from the testing set (top), the cluster-based GP1 (middle) and NB1 (bottom) predictions.

5   Conclusion

The presented chapter describes the prediction of the count variable with overdispersed realizations with the help of the five Poisson-related cluster-based models and compares them experimentally with the Poisson and negative binomial regressions. The cluster-based models are constructed using the relationship of the count variable with multi-modal multivariate Gaussian observations, using their cluster-based discretization. For the discretization, recursive algorithms of the Bayesian mixture estimation theory are applied. Experiments show that the cluster-based models excepting one of them demonstrate improvements in the prediction in the comparison with theoretical counterparts. In practice, the cluster-based Generalized Poisson models seem to be a balanced choice for predicting the considered type of data.

One of the main contributions of the proposed approach consists in the use of real-time continuous data to predict the target count variable described either by the Poisson or Poisson-related distributions. This advantage can be beneficial in specific applications, especially in view of its extension to other count data distributions, which do not have general conditional form.

Acknowledgements. This work was supported by the project Arrowhead Tools, the project number ECSEL 826452 and MSMT 8A19009. The authors would also like to thank Olga Cinková for her brilliant help and fruitful discussions during preparing this manuscript.

References

1. Guenni, L.B.: Poisson Distribution and Its Application in Statistics. In: Lovric, M. (ed.) International Encyclopedia of Statistical Science, pp. 1071–1072. Springer, Heidelberg (2011). https://doi.org/10.1007/978-3-642-04898-2_448
2. Jaggia, S., Kelly, A.: Business Statistics: Communicating with Numbers, 3rd edn. McGraw-Hill Education, New York (2018)
3. Doane, D., Seward, L.: Applied Statistics in Business and Economics, 3rd edn. Mcgraw-Hill, New York (2010)
4. Donnelly, R., Jr.: Business Statistics, 3rd edn. Pearson, London (2019)
5. Anderson, D.R., Sweeney, D.J., Williams, T.A., Camm, J.D., Cochran, J.J.: Essentials of Modern Business Statistics with Microsoft Office Excel (Book Only), 7th edn. Cengage Learning, Boston (2017)
6. Petrouš, M., Suzdaleva, E., Nagy, I.: Modeling of passenger demand using mixture of Poisson components. In: Gusikhin, O., Madani K., Zaytoon J. (eds.) Proceedings of the 16th International Conference on Informatics in Control, Automation and Robotics (ICINCO 2019), pp. 617–624 (2019)
7. Sørensen, Å.L., Lindberg, K.B., Sartori, I., Andresen, I.: Analysis of residential EV energy flexibility potential based on real-world charging reports and smart meter data. Energy Build. **241**, 110923 (2021)
8. Heeringa, S.G., West, B.T., Berglung, P.A.: Applied Survey Data Analysis. Chapman & Hall/CRC, Boca Raton (2010)
9. Falissard, B.: Analysis of Questionnaire Data with R. Chapman & Hall/CRC, Boca Raton (2012)

10. Armstrong, B.G., Gasparrini, A., Tobias, A.: Conditional Poisson models: a flexible alternative to conditional logistic case cross-over analysis. BMC Med. Res. Methodol. **14**, 122–128 (2014)

11. Agresti, A.: An Introduction to Categorical Data Analysis, 3rd edn. Wiley, Hoboken; New Jersey (2018)

12. Long, J.S., Freese, J.: Regression Models for Categorical Dependent Variables Using Stata, 3rd edn. Stata Press, College Station (2014)

13. Diallo, A.O., Diop, A., Dupuy, J.-F.: Analysis of multinomial counts with joint zero-inflation, with an application to health economics. J. Stat. Plan. Inference **194**, 85–105 (2018)

14. Agresti, A.: Categorical Data Analysis, 3rd edn. Wiley, Hoboken; New Jersey (2012)

15. Haykin, S.: Neural Networks: A Comprehensive Foundation. Macmillan, New York (1994)

16. Congdon, P.: Bayesian Models for Categorical Data. Wiley, Hoboken (2005)

17. Lim, H.K., Li, W.K., Yu, P.L.H.: Zero-inflated Poisson regression mixture model. Comput. Stat. Data Anal. **71**, 151–158 (2014)

18. Počuča, N., Jevtić, P., McNicholas, P.D., Miljkovic, T.: Modeling frequency and severity of claims with the zero-inflated generalized cluster-weighted models. Insur. Math. Econ. **94**, 79–93 (2020)

19. Perrakis, K., Karlis, D., Cools, M., Janssens, D.: Bayesian inference for transportation origin-destination matrices: the Poisson-inverse Gaussian and other Poisson mixtures. J. R. Stat. Soc. A. Stat. Soc. **178**, 271–296 (2015)

20. Yu, J., Gwak, J., Jeon, M.: Gaussian-Poisson mixture model for anomaly detection of crowd behaviour. In: Proceedings of 2016 International Conference on Control, Automation and Information Sciences (ICCAIS), pp. 106–111 (2016). https://doi.org/10.1109/ICCAIS.2016.7822444

21. Zha, L., Lord, D., Zou, Y.: The Poisson inverse Gaussian (PIG) generalized linear regression model for analyzing motor vehicle crash data. J. Transp. Saf. Secur. **8**, 18–35 (2016)

22. Silva, A., Rothstein, S.J., McNicholas, P.D., Subedi, S.: A multivariate Poisson-log normal mixture model for clustering transcriptome sequencing data. BMC Bioinform. **20**, 394 (2019)

23. Gupta, M.R., Chen, Y.: Theory and Use of the EM Method. (Foundations and Trends(r) in Signal Processing). Now Publishers Inc., Norwell (2011)

24. Li, Y., Sha, Y., Zhao, R.: Poisson prediction of the loss of teachers in high schools. In: Proceedings of 2010 International Conference on Multimedia Technology, Ningbo, China, pp. 1–3 (2010). https://doi.org/10.1109/ICMULT.2010.5629866

25. Bejleri, V., Nandram, B.: Bayesian and frequentist prediction limits for the Poisson distribution. Commun. Stat. Theory Methods **47**(17), 4254–4271 (2018)

26. Petrouš, M., Uglickich, E.: Modeling of mixed data for Poisson prediction. In: Proceedings of IEEE 14th International Symposium on Applied Computational Intelligence and Informatics (SACI 2020), Timisoara, RO, pp. 77–82 (2020). https://doi.org/10.1109/SACI49304.2020.9118836

27. Uglickich, E., Nagy, I., Petrouš, M.: Prediction of multimodal poisson variable using discretization of gaussian data. In: Gusikhin, O., Nijmeijer, H., Madani, K. (eds.) Proceedings of the 18th International Conference on Informatics in Control, Automation and Robotics (ICINCO 2021), pp. 600–608 (2021). https://doi.org/10.5220/0010575006000608

28. Gupta, A., Mehrotra, K., Mohan, C.K.: A clustering-based discretization for supervised learning. Stat. Probab. Lett. **80**(9–10), 816–824 (2010)

29. Kianmehr, K., Alshalalfa, M., Alhajj, R.: Fuzzy clustering-based discretization for gene expression classification. Knowl. Inf. Syst. **24**, 441–465 (2010)

30. Dash, R., Paramguru, R., Dash, R.: Comparative analysis of supervised and unsupervised discretization techniques. Int. J. Adv. Sci. Technol. **2**(3), 29–37 (2011)

31. Sriwanna, K., Boongoen, T., Iam-On, N.: Graph clustering-based discretization approach to microarray data. Knowl. Inf. Syst. **60**, 879–906 (2019)
32. Kárný, M., et al.: Optimized Bayesian Dynamic Advising: Theory and Algorithms. Springer, London (2006). https://doi.org/10.1007/1-84628-254-3
33. Nagy, I., Suzdaleva, E.: Algorithms and Programs of Dynamic Mixture Estimation. Unified Approach to Different Types of Components. SpringerBriefs in Statistics, Springer, Heidelberg (2017). https://doi.org/10.1007/978-3-319-64671-8
34. Lambert, D.: Zero-Inflated Poisson regression, with an application to defects in manufacturing. Technometrics **34**(1), 1–14 (1992). https://doi.org/10.2307/1269547
35. Zhang, H., Liu, Y., Li, B.: Notes on discrete compound Poisson model with applications to risk theory. Insur. Math. Econ. **59**, 325–336 (2014)
36. Best, D.J., Rayner, J.C.W., Thas, O.: Goodness of fit for the zero-truncated Poisson distribution. J. Stat. Comput. Simul. **77**(7), 585–591 (2007)
37. Yadav, B., et al.: Can Generalized Poisson model replace any other count data models? An evaluation. Clin. Epidemiology Glob. Health **11**, 100774 (2021)
38. Consul, P.C., Famoye, F.: Generalized Poisson regression model. Commun. Stat. Theor. Methods **21**, 89–109 (1992)
39. Hilbe, J.M.: Negative Binomial Regression. Cambridge University Press, Cambridge (2011)
40. Roy, D.: Discrete Rayleigh distribution. IEEE Trans. Reliab. **53**(2), 255–260 (2004)
41. Peterka, V.: Bayesian system identification. In: Eykhoff, P. (ed.) Trends and Progress in System Identification, pp. 239–304. Pergamon Press, Oxford (1981)
42. Gelman, A., Carlin, J.B., Stern, H.S., Dunson, D.B., Vehtari, A., Rubin, D.B.: Bayesian Data Analysis (Chapman & Hall/CRC Texts in Statistical Science), 3rd edn. Chapman and Hall/CRC, Boca Raton (2013)
43. Kárný, M., Kadlec, J., Sutanto, E.L.: Quasi-Bayes estimation applied to normal mixture. In: Rojíček, J., Valečková, M., Kárný, M., Warwick, K. (eds.) Preprints of the 3rd European IEEE Workshop on Computer-Intensive Methods in Control and Data Processing, CZ, Prague, pp. 77–82 (1998)
44. Cohen, A.C.: Estimating the parameter in a conditional Poisson distribution. Biometrics **16**(2), 203–211 (1960)
45. Consul, P.C.: Generalized Poisson Distributions: Properties and Applications. Marcel Dekker, New York (1989)
46. Date, S.: Generalized Poisson Regression for real world datasets. https://towardsdatascience.com. Accessed 4 Apr 2020
47. Siddiqui, M. M.: Statistical inference for Rayleigh distributions. J. Res. Natl. Bureau Stand. Sect. D Radio Sci. **68D**(9), 1007 (1964)
48. Date, S.: Negative Binomial Regression: A Step by Step Guide. https://towardsdatascience.com. Accessed 6 Oct 2019
49. Lewis, D.D.: Naive (Bayes) at forty: the independence assumption in information retrieval. In: Nédellec, C., Rouveirol, C. (eds.) ECML 1998. LNCS, vol. 1398, pp. 4–15. Springer, Heidelberg (1998). https://doi.org/10.1007/BFb0026666

Model-Based Optimization of Vaccination Strategies in Different Phases of Pandemic Virus Spread

Zonglin Liu[(✉)], Muhammed Omayrat, and Olaf Stursberg

Control and System Theory, Department of Electrical Engineering and Computer Science, University of Kassel, Kassel, Germany
{z.Liu,stursberg}@uni-kassel.de

Abstract. This paper aims at applying optimal control principles to investigate optimal vaccination strategies in different phases of a pandemic. Background of the study is that many countries have started their vaccination procedures against the COVID-19 disease in the beginning of 2021, but supply shortages for the vaccines prevented that everyone could be vaccinated immediately. At the beginning of 2022, in contrast, the vaccine supply was ample, but the effectiveness of different existing vaccines to avoid infection by new virus variants was in doubt, as well as the acceptance of booster doses decreased over time. To account for these effects, two formulations of optimization tasks based on different epidemic models are proposed in this paper. The solution of these tasks determines optimal distribution strategies for available vaccines, and optimized vaccination schemes to reduce the need of booster doses for later phase. Effectiveness of these strategies compared with other popular strategies (as applied in practice) is demonstrated through a series of simulations

Keywords: Epidemic modeling · Infection control · Optimal control · Epidemics · Vaccination

1 Introduction

The outbreak and rapid spread of COVID-19 in 2020 affected the life of almost everyone on the planet. Fortunately, the successful development and deployment of vaccines in the beginning of 2021 nurtured the hope that life may return to normal rather quickly, but production shortages at that time led to the situation that only a small share of the population could be vaccinated during the first half of 2021. Thus, the question of how to control the vaccination process subject to the given capacity constraints turned out to be crucial to mitigate the pandemic. One year later, in the beginning of 2022, after almost a whole year

© The Author(s), under exclusive license to Springer Nature Switzerland AG 2023
O. Gusikhin et al. (Eds.): ICINCO 2021, LNEE 1006, pp. 185–208, 2023.
https://doi.org/10.1007/978-3-031-26474-0_10

of mass vaccinations, many countries decided to relax all intervention rules in order to fully resume socio-economic activity. This decision was made although the new virus variant called *omicron*, which showed higher transmission rates but lower risks of hospitalization, was quickly spreading over the globe. This led to new (and still unresolved) questions for the continuation of the vaccination process, such as whether booster doses are useful for people of all ages, whether vaccines targeted to the new variant should be developed, or if one should simply regard the new variant as a *natural vaccine*. These questions, together with the problem of efficient vaccine distribution problem over the age groups in an initial phase of vaccination, are not specific for COVID-19, but must be considered relevant and central to any future virus pandemic. The goal of this paper is, accordingly, to provide answers by modeling the vaccination task as optimal control problems and solving these problems to determine optimal vaccination strategies for different phases of the pandemic. Vaccination strategies proposed in literature before the outbreak of COVID-19 include (but are not limited to) the uniform strategy [26], the targeted strategy [19], the random strategy [33], and the acquaintance strategy [6]. The work in [29] considered optimal vaccination strategies by taking into account the vaccination costs. The epidemic model used in this work is the one termed *susceptible-infected-susceptible* (SIS) model, see [10, 15] for more details. Also based on the SIS model, the work in [27] followed a similar pattern to develop an optimal vaccination strategy by taking into account additional constraints, while the work in [31] focused on how to exploit the topological structure of the virus spreading network to eliminate the virus in case of vaccine shortages. (The resulting strategy was then tested on a model for the SARS transmission in 2003). The work by [12] investigated how to efficiently distribute available medicine to minimize the overall infection. Unlike vaccines, which reduce the individual infection rate, the medicine in that work is assumed to increase the individual curing rate. The work in [16] proposed a robust vaccination strategy considering the case that many infections cannot be detected in due time. Instead of using the SIS model, a *susceptible-infected-recovered* (SIR) model was adopted there. After the outbreak of COVID-19, the authors in [23] have shown that vaccinating older age groups first is an optimal strategy for minimizing future deaths in the UK. Note that this conclusion on optimality is based on the reproduction number of the virus, instead of solving optimization problems as in this paper. In [9], a similar conclusion has also been made by comparing COVID-19 with the spread of the influenza virus. In the technical report [1], the authors formulated an optimal control problem to decide on the optimal vaccination strategy, and the resulting strategy was compared with random strategies, as well as an *older first strategy*. It should be noticed that for the two investigations in [23] and [1], the SIS model, or the SIR model respectively, was extended to include different types of the health status of individuals infected with COVID-19, such as exposed, symptomatic, or asymptomatic. Since an accurate model is crucial for the determination of reasonable vaccination strategies, the work in [5] focussed on real-time updating of the model with respect to the network underlying the virus spreading, such that individuals with high-degree of interaction or high-centrality can be identified and vaccinated. In the work mentioned above, vaccination strategies are determined each for one particular pair of epidemic model and vaccination goal. The observations for the COVID-19 pandemic in 2021 to 2022 have shown, however, that with the progress of the vaccination procedures as well as the mutations of the virus, both the epidemic model and the vaccination goal change over time. Typically, for the beginning phase of the vaccination in 2021, the vaccine was assumed to succeed in preventing individuals from infection by the original variant of the virus, and the risk of reinfections was (almost) ignored. The vaccination goal at that time thus was to reduce the overall infection. However, during the later phase in beginning of 2022, reinfections occurred more frequently according to [13,14], and a larger share of cases was observed in which the vaccines could not prevent infections with the newer Omicron variant. The vaccination goal in this phase was thus shifted to minimize the number of patients getting into critical stages of infection, see [25]. With these two phases of a pandemic in mind, the present paper aims at proposing schemes to compute optimized vaccination strategies for each phase and goal by using techniques from the field of optimal control. In the first part of the paper, referring to the first phase, a review of existing work on COVID-19 modeling is provided, and the epidemic model proposed in [11] is extended by including the vaccination process in Sect. 2. Based on this model, optimal control problems are formulated and solved in order to determine optimal vaccination strategies in Sect. 3. The solution takes into account different age groups of the population and heterogeneity with respect to infection and curing rates, as well as different contact situations and vaccine shortages in this phase. In Sect. 4, the computed strategies are simulated exemplarily for a midsize city and are compared to alternative strategies, as partly shown already in the own previous work [20]. In the second part of this paper, which refers to the second phase, first a modified epidemic model is proposed in Sect. 5., accounting for limited effectiveness of vaccines for virus variants. The corresponding optimal control problems to determine vaccination strategies for this phase are formulated in Sect. 6, and the solution process is sketched. The resulting strategies are compared to other strategies by numeric simulation in Sect. 7, before the paper is concluded in Sect. 8.

2 Epidemic Model Considering Partially Ineffective Vaccines

Most literature on epidemic modeling (without vaccination) prefer to use the SIS based model to describe the spread of a virus due to its simple structure, since only two states of any individual, susceptible and infected, are considered. After the outbreak of the COVID-19 pandemic, several alternative models have been proposed by different research groups taking into account the characteristics of this disease. Apart from those reviewed in the last section, the work in [21] extended the SIS model by taking into account topological aspects in terms of the transfer of infections between neighbored and connected regions. Likewise, the authors of [8] also considered the geography structure in Italy and extended the SIS model by four more states, namely quarantined, hospitalized, recovered, and deceased. In the work of [24], even more COVID-19 related states are covered and the influence of super-spreading individuals is also studied in this paper. The work in [32] further considered the effects of pre-existing illnesses on the spread of COVID-19 in Brazil and Spain. Among these efforts, the SIDARTHE model, which was proposed in [11] and is shown by the part marked in black in Fig. 1, has been adopted in different subsequent work, see [7,17,18,22,28] for example, as it covers more COVID-19 related states than any previous model. It distinguishes the following health status: being susceptible (S_i), infected (I_i), diagnosed (D_i), ailing (A_i), recognized (R_i), threatened (T_i), healed (H_i), and extinct (E_i). For a given population, these states of the SIDARTHE model represent the percentage of persons with different health status and the changes of these percentages is modeled stochastically by a continuous-time Markov process. Comparing with the original model in [11], the first extension of the present paper is to introduce to a set of n groups of persons, where the index $i \in N = \{1, \cdots, n\}$ refers to one of these groups (as in [20]). This extension will allow later to refer to a certain age-group of the population, in order to distinguish the heterogeneous mortality and curing rates of these groups.

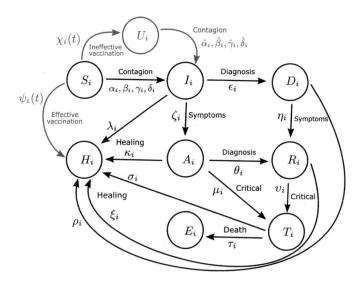

Fig. 1. The states and transitions in black represent the original SIDARTHE model, while the parts in red are newly included to represent the vaccination process, as well as are the indices i to model different groups of persons. Source [20].

Although the SIDARTHE model can well represent the development of the pandemic, vaccines at the time it was proposed in March 2020 were still not yet in sight. As noticed from the black part in Fig. 1, the health status cannot directly transfer from S_i to H_i without being infected (the case of reinfection was excluded at that time). To model the vaccination process in early 2021, i.e., the first phase under consideration in this paper, the SIDARTHE model is extended to incorporate an additional state U_i referring to the percentage of people in group i, for whom the vaccine is ineffective, as well as three new transitions $S_i \rightarrow H_i$, $S_i \rightarrow U_i$, and $U_i \rightarrow I_i$ (see the red part in Fig. 1). The new model (first proposed in the own previous work [20]) now also covers the following effects:

- the transition from S_i to H_i, which represents a successful vaccination step with rate $\psi_i(t) \in \mathbb{R}^{\geq 0}$. The rates $\psi_i(t)$ will be used as inputs determining the vaccination strategy to be optimized in the coming section;
- the transition from S_i to U_i, which represents an ineffective vaccination step with rate $\chi_i(t) \in \mathbb{R}^{\geq 0}$, where $\chi_i(t)$ is assumed to be proportional to $\psi_i(t)$, i.e., $\chi_i(t) = q \cdot \psi_i(t)$, $q \in \mathbb{R}^{>0}$;
- the transition from U_i to I_i, which represents an infection step with respective contagion parameters $\hat{\alpha}_i$, $\hat{\beta}_i$, $\hat{\gamma}_i$, and $\hat{\delta}_i$. Note that the labeling of this transition with four rates refers to the shares that a person in state U_i gets infected by contact with a person in state I_i, D_i, A_i or R_i respectively (the same principle underlies the assignment of four parameters to the transition from S_i to I_i).

In order to model the evolution of the probability distribution over the discrete states for any group $i \in N$, the following nonlinear dynamics first proposed in [20] is selected:

$$\dot{S}_i(t) = -S_i(t) \sum_{j \in N} W_{ij}(\alpha_i I_j(t) + \beta_i D_j(t) + \gamma_i A_j(t) + \delta_i R_j(t)) - (\chi_i(t)$$
$$+ \psi_i(t))S_i(t) \tag{1}$$

$$\dot{U}_i(t) = -U_i(t) \sum_{j \in N} W_{ij}(\hat{\alpha}_i I_j(t) + \hat{\beta}_i D_j(t) + \hat{\gamma}_i A_j(t) + \hat{\delta}_i R_j(t)) + \chi_i(t)S_i(t) \tag{2}$$

$$\dot{I}_i(t) = S_i(t) \sum_{j \in N} W_{ij}(\alpha_i I_j(t) + \beta_i D_j(t) + \gamma_i A_j(t) + \delta_i R_j(t)) - (\epsilon_i + \zeta_i$$
$$+ \lambda_i)I_i(t) + U_i(t) \sum_{j \in N} W_{ij}(\hat{\alpha}_i I_j(t) + \hat{\beta}_i D_j(t) + \hat{\gamma}_i A_j(t) + \hat{\delta}_i R_j(t)) \tag{3}$$

$$\dot{D}_i(t) = \epsilon_i I_i(t) - (\eta_i + \rho_i)D_i(t) \tag{4}$$

$$\dot{A}_i(t) = \zeta_i I_i(t) - (\theta_i + \mu_i + \kappa_i)A_i(t) \tag{5}$$

$$\dot{R}_i(t) = \eta_i D_i(t) + \theta_i A_i(t) - (v_i + \xi_i)R_i(t) \tag{6}$$

$$\dot{T}_i(t) = \mu_i A_i(t) + v_i R_i(t) - (\sigma_i + \tau_i)T_i(t) \tag{7}$$

$$\dot{H}_i(t) = \lambda_i I_i(t) + \rho_i D_i(t) + \kappa_i A_i(t) + \xi_i R_i(t) + \sigma_i T_i(t) + \psi_i(t)S_i(t) \tag{8}$$

$$\dot{E}_i(t) = \tau_i T_i(t) \tag{9}$$

Note that the model also satisfies $\dot{S}_i(t) + \dot{U}_i(t) + \dot{I}_i(t) + \dot{D}_i(t) + \dot{A}_i(t) + \dot{R}_i(t) + \dot{T}_i(t) + \dot{H}_i(t) + \dot{E}_i(t) = 0$ at any time t for any choice of parameters. These parameters are denoted by α_i, β_i, γ_i, δ_i, $\hat{\alpha}_i$, $\hat{\beta}_i$, $\hat{\gamma}_i$, $\hat{\delta}_i$, ϵ_i, ζ_i, λ_i, η_i, θ_i, v_i, ξ_i, τ_i, ρ_i, σ_i, μ_i, and one can estimate their values according to statistical reports from, e.g., the Robert-Koch Institute in Germany, or the Center for Disease Control and Prevention in the US, by using regression methods. In this paper, their values are assumed to be known[1] for all groups $i \in N$. Note further that the contagion parameters assigned to $U_i \rightarrow I_i$ are different from those assigned to $S_i \rightarrow I_i$ to account for the possibility that a vaccinated person, which is unaware of the fact that the vaccination was not successful, will likely follow more relaxed contact patterns, compared to persons that still wait to be vaccinated. In order refer to the above model in brief, let it be denoted by:

$$\dot{x}_i(t) = f(x_i(t), \psi_i(t)), \; x_i(t_0) = x_{i,0} \tag{10}$$

with state vector $x_i(t) := (S_i(t), U_i(t), I_i(t), D_i(t), A_i(t), R_i(t), T_i(t), H_i(t), E_i(t))$, and the initial state $x_{i,0}$. Furthermore, the model aims at considering the spread of the virus across different age groups of the population. In order to formalize the contacts between different age groups, and thus the possibility that infections occur across the groups, an undirected graph $G = \{N, E\}$ is set up as in [20]. The set N of nodes here models the different age groups, with indices $i \in N$ as before. The set E of undirected edges models the contact between a pair of groups, and for any edge $e_{i,j} \in E$ between the groups with indices i and j, a positive weight $W_{ij} \in \mathbb{R}^{>0}$ is assigned, representing that infections in group j can affect the infection in group i (and vice versa), see also (1)–(3) in the model. More precisely, a weight W_{ij} models the average share of time that a person from group i spends with a person from group j. Accordingly, a self-loop transition with weight W_{ii} is introduced for any node, and the matrix $W \in \mathbb{R}^{n \times n}$ is chosen as doubly stochastic matrix, i.e., $\sum_{j \in N} W_{ij} = 1$ for all $i \in N$, and $\sum_{i \in N} W_{ij} = 1$ for all $j \in N$.

It should be obvious that the considered epidemic model is a direct extension of the SIDARTHE model. The more detailed representation of the effects of vaccination are a prerequisite to study (in the upcoming section) how vaccination strategies under consideration of shortages of vaccines can be designed based on model optimization.

3   Optimized Vaccination Strategies for Limited Amount of Vaccines

Based on the model (10), optimized vaccination strategies by using principles of optimal control principles are determined in this section. The underlying idea is based on the observation, that the infection has, on average, more severe effects for the older generations, so that they should be prioritized by vaccination. But the younger generations, on the other hand, have higher contact rates among

---

[1] Such estimation, of course, will result in model uncertainty. In order to consider e.g. parameter intervals, extensions to techniques of robust predictive control could be employed, see e.g. [3], but this is outside of the scope of the present paper.

each other, thus leading to higher risk of infection if they are not vaccinated in time. To address this situation, the population is divided into n groups according to their age, and for any age-group with index $i \in N = \{1, \cdots, n\}$, let its share of the total population be denoted by v_i, and $\sum_{i \in N} v_i = 1$. The vaccination of different age groups (represented by the rates $\psi_i(t)$ and $\chi_i(t)$ in Fig. 1) as well as contacts between age groups (modelled by the weights W_{ij}) can be expected to have significant impact on the evolution of the epidemic. In order to set up, an optimal control formulation, first let an initial time t_0 be given as well as a time interval $[t_0, t_0 + H]$, where H is the number of days for the considered phase of the vaccination process. Assume further that the vaccination strategy can only be adjusted every T days of the horizon, leading to totally $\frac{H}{T}$ decision steps. Let $k \in \{0, 1, \cdots, \frac{H}{T} - 1\}$ index these steps. Once the strategy is determined for k, it is held constant for the coming T days. The vaccination rate applied to any age-group $i \in N$ on the interval $[k \cdot T, (k+1) \cdot T[$ is referred to by $\psi_{i,k}$. To model the problem of vaccine shortages, the amount of vaccine available for step k over all groups i is denoted by $\Psi_{max,k}$, leading to the following constraint [20]:

$$\sum_{i \in N} \mathcal{P} \cdot v_i \cdot (\psi_{i,k} + \chi_{i,k}) \leq \Psi_{max,k}, \tag{11}$$

with a population size $\mathcal{P}$, and $\mathcal{P} \cdot v_i$ representing the size of age-group i. Corresponding to the relation between $\chi_i(t)$ and $\psi_i(t)$ as mentioned in Sect. 2, it follows that: $\chi_{i,k} := q \cdot \psi_{i,k}$. Given the constraint (11), the task henceforth is to optimize the vaccination strategy $\psi := (\psi_0, \psi_1, \ldots, \psi_{\frac{H}{T}-1})$ with $\psi_k = (\psi_{1,k}, \ldots, \psi_{n,k})^T$ for all age-groups in all decision steps, such that the infected share of the population, and/or the number of fatalities is minimized. The infected share is determined by all persons assigned to the states I_i, D_i, A_i, R_i, T_i in Fig. 1, while the number of fatalities corresponds to the number of persons associated with the state E_i. To this end, the following cost function is proposed:

$$J_1(\psi) := \sum_{i \in N} \int_{t_0}^{t_0+H} \mathcal{P}v_i(c_1 \cdot (I_i(t) + D_i(t) + A_i(t) + R_i(t) + T_i(t)) + c_2 \cdot E_i(t))dt. \tag{12}$$

The parameters c_1 and c_2 (both positive) denote appropriate weights of the terms of the cost function. Of course, the function $J_1(\psi)$ may also be extended to additional cost terms to account for, such as the costs of testing or medical treatment. Likewise, additional constraints, such as limitations in available staff for vaccination could also be considered.

In order to determine an optimized vaccination strategy ψ^* for the first phase, the following optimization problem (which was first proposed in [20]) is defined according to the objectives and constraints above as well as the model (10):

Problem 1.

$$\min_{\psi} J_1(\psi) \tag{13}$$

s.t. for all $i \in N$ and given the graph G :

$$x_i(t_0) := x_{i,0}, \text{ dynamics (10), } for \ all \ t \in [t_0, t_0 + H], \tag{14}$$

$$for \ all \ k \in \{0, \ldots, \frac{H}{T} - 1\} :$$

$$\psi_{i,k} \geq 0, \tag{15}$$

$$\sum_{i \in N} \mathcal{P} \cdot v_i \cdot (1 + q)\psi_{i,k} \leq \Psi_{max,k}. \tag{16}$$

The optimized strategy ψ^* is then determined by solving this nonlinear continuous-time optimization problem and it is noticed that this problem cannot be locally solved by each group i. This is because the constraint (16) determines a coupling constraint among all groups and the dynamics of each group are also coupled due to the matrix W in (1)–(3). A possible centralized solution for Problem 1 is using multiple shooting methods, which then casts the original problem into a finite dimensional nonlinear programming problem by parameterizing the input and state space, see e.g. [2]. However, as Problem 1 is non-convex one may only obtain a sub-optimal strategy as solution, and this is also why it is referred to an optimized strategy rather than an optimal one.

4   Simulation for Limited Amount of Vaccines

To simulate the effectiveness of optimized vaccination strategies from Problem 1 and also to discuss a number of effects for the first phase of the vaccination, a midsize city with a population of $\mathcal{P} = 200,000$ is considered in the following simulations. According to a statistics of distribution of the population over the age groups $i \in N$ (in Germany for 2021), the values in Table 1 are assigned to the city.

Table 1. Distribution over age groups. Source [20].

Age	0–19	20–39	40–59	60–79	80+
v_i	18.5%	24.6%	28.8%	21.6%	6.5%

The estimated average time (in percentage) persons of one group spend with those of another is listed in Table 2, constituting the entries of matrix W.

Note that Table 2 also reflects certain intervention policies deployed in this phase in Germany, such as that visiting nursing homes was forbidden (such that the most senior group had to spent most of their time with its own), while schools were still open. The parameters contained in the dynamics (1)–(9) are chosen

Table 2. Contacts between member of different age-groups during the epidemic. Source [20].

Age	0–19	20–39	40–59	60–79	80+
0–19	31%	34%	25%	8%	2%
20–39	34%	45%	15%	5%	1%
40–59	25%	15%	30%	20%	10%
60–79	8%	5%	20%	50%	17%
80+	2%	1%	10%	17%	70%

similar to [11], except of the fact that the contagion parameters $\hat{\alpha}_i, \hat{\gamma}_i$ for the transition from $U_i \to I_i$ are chosen larger than for $S_i \to I_i$, and the effectiveness of the vaccine is assumed to be 90% for all age-groups, i.e., $\chi_i := 0.11\psi_i$, $i \in N$. The former aims at reflecting the effect that people are less cautious to the virus after being vaccinated, although the vaccination may be ineffective, while the latter can be extended to considering different effectiveness rates of the vaccine for different age-groups, or to different rates for different types of vaccines. The initial states for t_0 are chosen according to the pandemic situation of the city at the beginning of 2021. The considered horizon is $H = 90$ days, and the vaccination strategy can be adjusted every $T = 30$ days (thus, in total 3 decision steps). The weights c_1 and c_2 in $J_1(\psi)$ are selected, such that the main goal is to reduce deaths.

4.1   Uniform Vaccination Strategy

Before the optimized strategy is tested, a uniform vaccination strategy is first simulated for comparison reason: It is assumed that the availability of the vaccine is constant in all decision steps, i.e., $\Psi_{max,k} := \Psi_{max}$, $\forall k \in \{0, 1, 2\}$, and the $\psi_{i,k}$ in the uniform vaccination strategy satisfy: $\psi_{i,k} := \frac{\Psi_{max}}{\mathcal{P} \cdot (1+q)}$ for all $i \in N$. When adopting this strategy, the outcome as illustrated in Fig. 2 is obtained. It can be noticed that most of the deaths occur in the senior groups **60–79** and **80+**, although the number of infections in these groups (corresponding to the states in I_i, D_i, A_i, R_i, and T_i) are not the highest. This is mainly because of the high mortality rate of these two groups.

4.2   Optimized Vaccination Strategy

In the second test, an optimized vaccination strategy $\psi_{i,k}^*$ for each group is determined by solving Problem 1, see Fig. 3 for more details of this strategy.

By deploying this strategy, the resulting infections and deaths are illustrated in Fig. 4. Compared to the outcome of the uniform strategy in Fig. 2, more infected cases occur in the groups **20–39** and **40–59**, while less in the groups **60–79** and **80+**. The deaths for the two senior groups, on the contrary, are significantly reduced by using the optimized strategy, which leads to a decrease from 329 to 174, see Fig. 5.

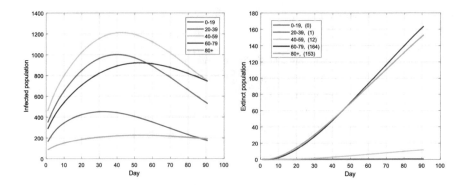

Fig. 2. Number of infections (left) and deaths (right) when applying the uniform strategy over $H = 90$ days. Source [20].

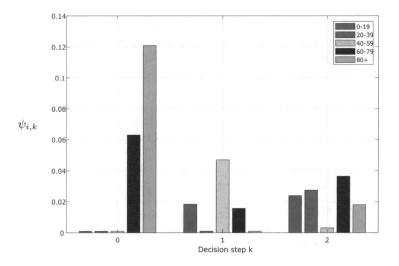

Fig. 3. The optimized vaccination strategy from Problem 1: in $k = 0$, the complete available vaccine is distributed to the groups **60–79** and **80+**; in $k = 1$, most of the vaccine is distributed to the group **40–59**, **60–79** and **0–19**; in the last step $k = 2$, the vaccine is distributed to all groups besides the group **40–59**. Source [20].

The optimized strategy is further compared to the popular strategy of first vaccinating the older generations, which has been adopted by governments in many countries, see Fig. 6. All available vaccines are first provided to the group **80+** in the steps $k = 0$ and $k = 1$ in this strategy, and then to the group **60–79** for $k = 2$. One can notice from the outcome, however, that only the number of deaths in the group **80+** is slightly reduced compared to the uniform strategy, while much more occur in the group **60–79**. This is due to the larger size of the latter group than the **80+** group, as well as due to the more frequent contacts of this group with junior groups (see Table 2). The immunity of the group **60–79**,

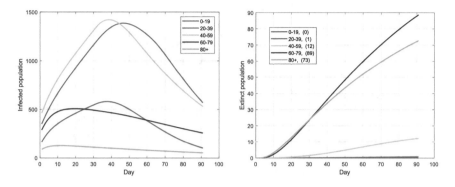

Fig. 4. Infections (left) and deaths (right) with the optimized strategy. Source [20].

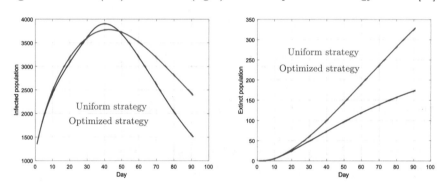

Fig. 5. Infections (left) and deaths (right) over all age-groups by adopting the uniform vaccination strategy and the optimized one. Source [20].

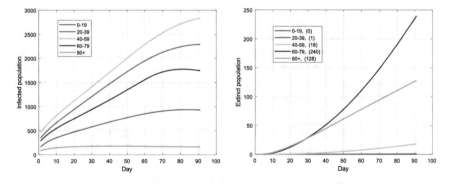

Fig. 6. Infections (left) and deaths (right) when applying the older-first strategy. Source [20].

in contrast, is only slightly better than for the **80+** group, but much worse than for the junior groups. Based on this simulation result, accordingly, it is not wise to ignore the group **60–79** in the first step of the vaccination process.

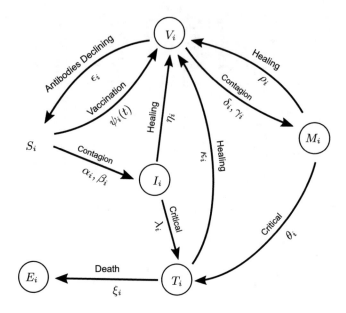

Fig. 7. Epidemic model tailored to a phase with reduced efficacy of vaccination.

5   Epidemic Modeling for Reduced Motivation for Vaccination and New Virus Variants

At the beginning of 2022, i.e. after a whole year of mass vaccinations, a large share of the population in Europe has been vaccinated at least twice, but the pace of vaccinating persons not been vaccinated before as well as that of applying booster doses has fallen significantly. This is partly due to reduced motivation of individuals for taking additional shots of vaccine, whereas the amount of antibodies was found to decline relatively quickly over time [30]. In addition, some data indicates that the existing vaccines provide only moderate protection against infections caused by the Omicron variant, while severe disease occur very seldom [4]. These facts motivate to adapt the vaccination strategy to the new situation.

Therefore, a new epidemic model is proposed for this phase, see Fig. 7. The new model only includes six states related to the health status, namely: lack of antibodies (S_i), sufficient amount of antibodies (V_i), being infected in case of a lack of antibodies (I_i), being infected even despite a sufficient amount of antibodies (M_i), critical (T_i), and extinct (E_i). The transitions between different states are defined as follow:

- $S_i \rightarrow V_i$: the antibody level can be increased through vaccination; the corresponding vaccination rates $\psi_i(t)$ will be further used as inputs to be optimized;
- $V_i \rightarrow S_i$: the antibody declines with rate ϵ_i over time;

- $S_i \rightarrow I_i$ and $V_i \rightarrow M_i$: persons with lack of antibodies (in S_i) or with sufficient antibodies (in V_i) may both be infected and thus be transferred to I_i and M_i respectively, but with different infection rates (α_i, β_i) and (δ_i, γ_i).
- $I_i \rightarrow T_i$ and $M_i \rightarrow T_i$: patients in either I_i or M_i may both need to be hospitalized if severe symptoms occur; studies have shown that the risk fur such symptoms is much higher for those without sufficient amount of antibodies [4], i.e., the rate λ_i is much higher than θ_i;
- $I_i \rightarrow V_i$, $M_i \rightarrow V_i$, and $T_i \rightarrow V_i$: patients in I_i, M_i, and T_i may eventually recover from infection, but with different rates η_i, ρ_i, and κ_i. In addition, since the antibody level can also be increased through infection, V_i is selected as the target state for these transitions;
- $E_i \rightarrow V_i$: patients in critical condition may eventually die with rate ξ_i.

Compared to the model in Fig. 1, the states such as ailing or diagnosed are excluded from the new model. Also the vaccination here can only increase the antibody level, instead of preventing infection. Similar to (1)–(9), the evolution of the probability distribution over these states can be described for any group $i \in N$ using the following nonlinear dynamics:

$$\dot{S}_i(t) = -S_i(t) \sum_{j \in N} W_{ij}(\alpha_i I_j(t) + \beta_i G_j(t)) - \psi_i(t)S_i(t) + \epsilon_i V_i(t) \tag{17}$$

$$\dot{V}_i(t) = -V_i(t) \sum_{j \in N} W_{ij}(\delta_i I_j(t) + \gamma_i G_j(t)) + \psi_i(t)S_i(t) + \eta_i I_i(t) + \kappa_i T_i(t)$$

$$+ \rho_i M_i(t) - \epsilon_i V_i(t) \tag{18}$$

$$\dot{I}_i(t) = S_i(t) \sum_{j \in N} W_{ij}(\alpha_i I_j(t) + \beta_i G_j(t)) - (\eta_i + \lambda_i)I_i(t) \tag{19}$$

$$\dot{M}_i(t) = V_i(t) \sum_{j \in N} W_{ij}(\delta_i I_j(t) + \gamma_i G_j(t)) - (\rho_i + \theta_i)M_i(t) \tag{20}$$

$$\dot{T}_i(t) = \lambda_i I_i(t) + \theta_i M_i(t) - (\kappa_i + \xi_i)T_i(t) \tag{21}$$

$$\dot{E}_i(t) = \xi_i T_i(t) \tag{22}$$

and the weights W_{ij} follow the same definition as before. This model is referred to in the following brief form:

$$\dot{x}_i(t) = f_2(x_i(t), \psi_i(t)), \quad x_i(t_0) = x_{i,0} \tag{23}$$

with state vector $x_i(t) := (S_i(t), V_i(t), I_i(t), M_i(t), T_i(t), E_i(t))$.

6   Optimized Vaccination Strategies for New Virus Variants

According to the COVID-19 vaccine report of the WHO (World Health Organization) [25], the vaccination goal in this phase of the pandemic was to minimize the number of deaths and incidents of severe disease. Considering also the fact that the acceptance of additional booster doses is decreasing over time, the following form of the cost function is chosen for this phase:

$$J_2(\psi) := c_1 \cdot \sum_{i \in N} \int_{t_0}^{t_0+H} \mathcal{P}v_i \cdot T_i(t) \, dt + c_2 \cdot \sum_{k \in \{0, \cdots, \frac{H}{T}-1\}} \sum_{i \in N} \mathcal{P}v_i \cdot \psi_{i,k} \quad (24)$$

with positive weights c_1 and c_2. Note that only the number of patients in critical shape is considered in $J_2(\psi)$, since the transition from T_i (critical) to E_i (death) follows a constant rate ξ_i (thus minimizing T_i is equivalent to minimizing E_i). The second term in $J_2(\psi)$ refers to minimizing the overall need for vaccines. With respect to constraints of the optimization problem, one can assume that the number of patients in critical condition should not exceed an upper bound T_{max} (depending upon the hospital staff reserved for COVID-19 patients), i.e.:

$$\sum_{i \in N} \mathcal{P}v_i \cdot T_i(t) \le T_{max}, \text{ for all } t \in [t_0, t_0 + H]. \quad (25)$$

Henceforth, the following optimization problem is solved to determine an optimized vaccination strategy in this phase:

Problem 2.

$$\min_{\psi} J_2(\psi) \quad (26)$$

s.t. for all $i \in N$ and given G :

$$x_i(t_0) = x_{i,0}, \text{ dynamics (23)}, \text{ for all } t \in [t_0, t_0 + H], \quad (27)$$

$$\sum_{i \in N} \mathcal{P}v_i \cdot T_i(t) \le T_{max}, \text{ for all } t \in [t_0, t_0 + H], \quad (28)$$

$$\psi_{i,k} \ge 0, \text{ for all } k \in \{0, \ldots, \frac{H}{T} - 1\}. \quad (29)$$

Note that the Problems 1 and 2 belong to the same class of optimization problems and thus can be solved by the same method.

Remark 1. Apart from computing an optimized vaccination strategy, the model in Fig. 7 can also be extended to investigate how effective a new vaccine must be in order to quickly mitigate the pandemic. In this case, the rates δ_i, γ_i of the transition $V_i \to M_i$ and the rate ϵ_i of transition $V_i \to S_i$ are selected as optimization variables. Alternatively, one can also use this model to study the influence of new treatments developed for COVID-19 by adjusting the rate θ_i for $M_i \to T_i$. Both extensions will be illustrated in the following simulations.

7   Simulation for the Modified Model

In this section, the same city as before is considered and the age distribution in Table 1 and the matrix W in Table 2 are also directly adopted from Sec. 4

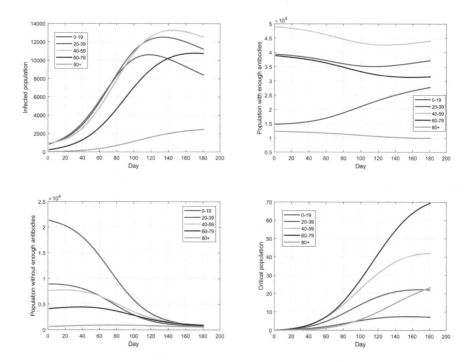

Fig. 8. By regarding the Omicron as the natural vaccine, the development of the pandemic over $H = 180$ days for all age groups is shown for: the infected population (top left), the population with sufficient antibodies (top right), the population lack of antibodies (down left), and the critical patients (down right).

(under the assumptions that the pandemic has not significantly affected the age structure and that the people were getting used to the contact routines). The initial states at t_0 are chosen according to the pandemic situation of the city at the beginning of 2022, i.e., around 3000 infections are present and the vaccination ratios for the five age groups (from young to old) are $[40\%, 80\%, 85\%, 90\%, 95\%]$. The critical and death cases are set to zero at t_0, to consider only new cases from there on. The transition rates are selected to approximate the pandemic at the time when Omicron has become the predominant variant. Typically, these rates (for all age groups) satisfy:

- The critical rate λ_i for $I_i \to T_i$ is 10 times higher than θ_i for $M_i \to T_i$;
- The healing rate ρ_i for $M_i \to V_i$ is twice as high than η_i for $I_i \to V_i$;
- The contagion rates δ_i, γ_i for $V_i \to M_i$ are lower than α_i, β_i for $S_i \to I_i$;
- The declining rate ϵ_i of the antibodies is selected such that it will take three months for individuals to transfer from V_i to S_i.

The simulation horizon is $H = 180$ days, and the vaccination strategy can now be adjusted every $T = 90$ days (i.e. two decision steps in total).

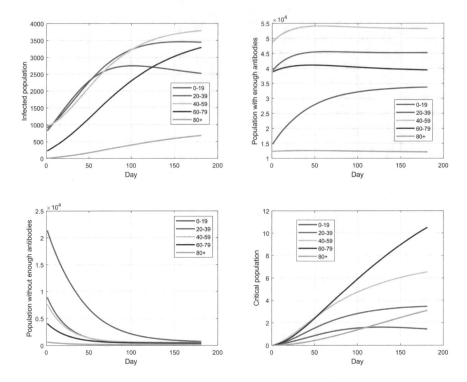

Fig. 9. Development of the pandemic when continuing the strategy of mass vaccinations as in the second half of 2021.

7.1   Omicron as a Natural Vaccine

Before addressing optimized vaccination strategies, first the case is evaluated the Omicron variant of the virus is considered as a natural vaccine. It is assumed that the vaccination procedure is stopped from t_0 on, i.e., $\psi_i(t) = 0$ for all $i \in N$ and for all $t \in [t_0, t_0 + H]$. The development of the pandemic in $H = 180$ days is illustrated in Fig. 9.

It can be observed that the cases of infection cases (in I_i and M_i) of all age groups quickly increase due to the high infectivity of Omicron, leading to around 25% of the whole population being infected at the same time after 120 days. The population with sufficient antibodies (V_i) also decreases in this period for all groups, except of the age group of **0–19**. This is because the latter group has a significantly lower vaccination ratio at t_0, as not all kids are recommended to be vaccinated. After 120 days, the population with sufficient antibodies starts to increase because most of them are recovered from infection. The cases of infection, in contrast, start to decrease after 120 days, while the number of patients in critical condition (T_i) has been constantly increasing over the entire horizon. The share of the population which with a lack of antibodies (S_i) decreases over the whole horizon, until a minimum value has been reached (this is because most

of the people are either infected or possess enough antibodies through infection).
Finally, a total number of 72 deaths results and the pandemic appears to slowly
come to an end (assuming that no new virus variants appear).

7.2   The Strategy of Mass Vaccination

In the second test, the effect of continuing mass vaccinations as in the second
half of 2021 is simulated.

All people in the city are assumed to receive successive booster doses
(although the vaccine can only provide protection against severe disease and
death) before their antibody levels reduce to a too low level. This strategy, of
course, implies that all stations and the staff for vaccination must be preserved,
leading to high economic costs. The outcome is illustrated in Fig. 8. Compared
to the result in Fig. 9, the mass vaccination strategy can indeed significantly
reduce both, the number of infections and of patients in critical condition. In this
respect, keeping the strategy of mass vaccinations provides an effective means to
mitigate the pandemic (before new vaccines targeted to Omicron are developed
and deployed).

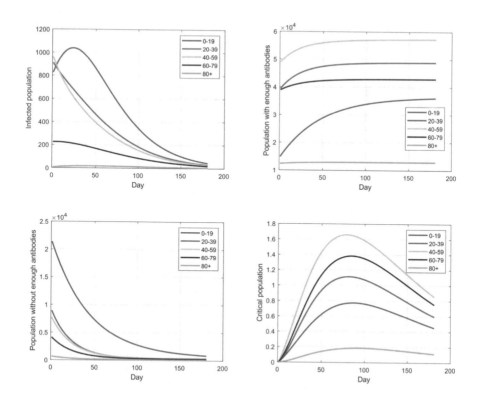

Fig. 10. Development of the pandemic by using new vaccine targeted to Omicron.

7.3   New Vaccine Targeted to Omicron

In the third test, it is assumed that a modified vaccine tailored to the Omicron variant becomes available and is immediately deployed. If the new vaccine can reduce the contagion rates δ_i, γ_i for $V_i \rightarrow M_i$ to 20% of the value of present vaccines, as well as decreasing the declining rate ϵ_i of antibodies to 50% of the previous rate, and if the same mass vaccination strategy is applied as in the last test, the result as illustrated in Fig. 10 is obtained. It shows that the new vaccine can almost immediately reduce the cases of infection (except for the group of **0–19**), while the number of patients in critical shape is also reduced to a very low level. In practice, however, the distribution of the new vaccine may again suffer supply shortages, as in the phase considered in Sect. 3. To this end, an optimized distribution plan of the new vaccine can once more be determined by solving Problem 1.

7.4   New Treatments for Infected Patients

In a further test, new treatments against severe cases of the diseases are assumed to be available. By using the new treatments to infected people, the critical rate λ_i and θ_i for transitions $I_i \rightarrow T_i$ and $M_i \rightarrow T_i$ can be reduced to 20% of their current values. It is shown in Fig. 11 that, if the vaccination procedure is stopped since t_0, and one purely relies on the new treatments against the pandemic, the

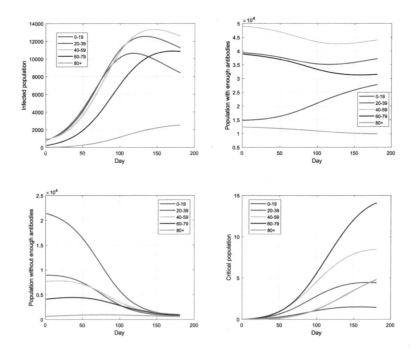

Fig. 11. Development of the pandemic when new treatments are deployed.

number of infections is almost the same as when regarding Omicron as a natural vaccine, see Fig. 9. Here, the number of patients in critical shape can be reduced to a relatively low level (similar to that of the mass vaccination strategy in Fig. 8).

7.5   Optimized Vaccination Strategy

Finally the case is considered that neither new vaccines targeted to Omicron nor new effective treatments are available, and that the acceptance of receiving further booster doses decreases over time, thus preventing further application of mass vaccinations. For this case, an optimized strategy aiming at reducing both, severe cases of the disease (as well as fatalities) and the overall amount of vaccination, is determined by solving Problem 2. The constraint (25) in Problem 2 is employed to achieve that the number of patients in critical condition does not exceed (an arbitrarily selected number of) 50 cases in any day of the horizon. For the cost function $J_2(\psi)$, the task of minimizing the need of vaccination and the number of patients in critical shape are assumed to be equally important. The outcome by deploying the obtained strategy is shown in Fig. 12. It can be noticed that the number of critical patients stays always below the specified upper bound (thus similar to the strategy of mass vaccination), while

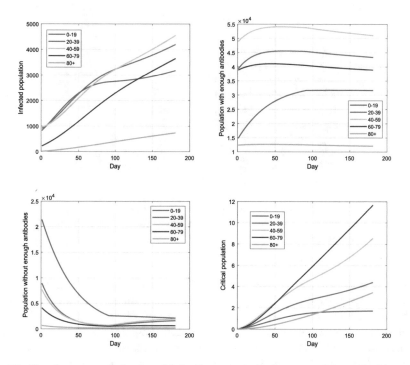

Fig. 12. Development of the pandemic if an optimized strategy from Problem 2 is deployed.

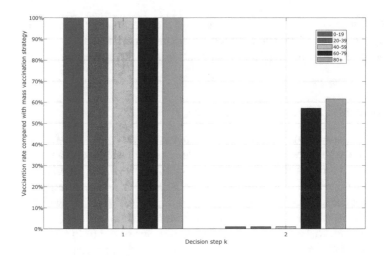

Fig. 13. Comparison between the mass vaccination strategy and the optimized strategy from Problem 2 in each decision step k.

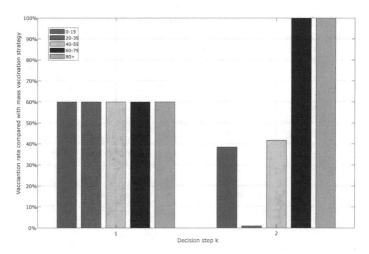

Fig. 14. In case only 60% of the vaccination rate in the first step of the strategy in Fig. 13 can be achieved, adjustment must be made for the second step (by solving Problem 2 once more).

the overall need for vaccination has been significantly reduced compared to the latter strategy, see Fig. 13. In the obtained strategy, mass vaccination is applied in the first decision step, while only the group of **60–79** and **80+** should be vaccinated in the second step (with a rate approximately 50% lower than that of mass vaccination). A further test underlies the assumption that only 60% of the desired rate in the first decision of the obtained strategy can be achieved

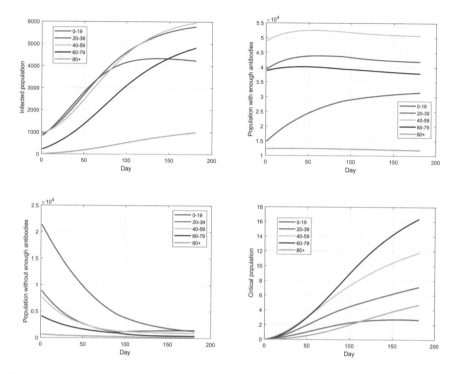

Fig. 15. Development of the pandemic when the strategy in Fig. 14 is deployed. Clearly, more critical cases result compared to Fig. 12.

(for all age groups), and the second decision is optimized by solving Problem 2. Figure 14 shows the optimized strategy and Fig. 15 the course of the pandemics for this case. Obviously, the two senior groups **60–79** and **80+** should be vaccinated to an as high extent as possible in the second step. These results, together with those shown in Fig. 9 and 8 may provide meaningful insight for regulators to determine reasonable vaccination strategies for situations as in the first half of 2022 (before new vaccines targeted to Omicron are deployed).

8 Conclusions

In this paper, vaccination strategies for two different phases of the COVID-19 pandemic have been determined according to principles of optimal control. For the first phase, referring to the beginning of 2021, the known SIDARTHE model has been first extended in different respects in order to be tailored to the study of vaccination procedures against the epidemic virus spread. The motivation of these extensions is the typical shortage of vaccines in this phase. To effectively combat the epidemic with a limited amount of vaccine, the solution of optimal control problems are applied to determine strategies for optimal distribution of the vaccine among different age-groups. For a second phase of the pandemic, referring to the beginning of 2022, another epidemic model is proposed to account for the new variant of the virus (Omicron) and a different situation for vaccination, namely that existing vaccines cannot prevent infection and that the use of booster doses is reduced. Based on this model, again optimized vaccination strategies are determined through the solution of a modified optimal control problem, meaning that only specific age groups are required to receive booster doses, while the vaccination goal of reducing deaths and severe disease can still be achieved.

For the situation that the transition rates or model parameters are only known within certain intervals, the optimal control problems in both phases could be modified into one of optimization under uncertainty, e.g. adopting techniques of robust optimal control – this is subject of future investigations. In addition, to account for the occurrence of new mutations, the solution of the optimal control problem could be recursively solved over time, leading to a strategy of model predictive control.

References

1. Angelov, G., Kovacevic, R., Stilianakis, N., Veliov, V.: Optimal vaccination strategies using a distributed epidemiological model applied to COVID-19 (2021)
2. Bock, H., Diehl, M., Leineweber, D., Schlöder, J.: A direct multiple shooting method for real-time optimization of nonlinear DAE processes. In: Allgöwer, F., Zheng, A. (eds.) Nonlinear Model Predictive Control, pp. 245–267. Springer, Basel (2000). https://doi.org/10.1007/978-3-0348-8407-5_14
3. Campo, P., Morari, M.: Robust model predictive control. In: American Control Conference, pp. 1021–1026. IEEE (1987)
4. Carreño, J., Alshammary, H., Tcheou, J., et al.: Activity of convalescent and vaccine serum against SARS-CoV-2 Omicron. Nature **602**(7898), 682–688 (2022)
5. Cheng, S., Pain, C., Guo, Y., Arcucci, R.: Real-time updating of dynamic social networks for COVID-19 vaccination strategies. arXiv preprint arXiv:2103.00485 (2021)
6. Cohen, R., Havlin, S., Ben-Avraham, D.: Efficient immunization strategies for computer networks and populations. Phys. Rev. Lett. **91**(24), 247901 (2003)
7. Cooper, I., Mondal, A., Antonopoulos, C.: A sir model assumption for the spread of COVID-19 in different communities. Chaos Solitons Fractals **139**, 110057 (2020)
8. Della, F., et al.: A network model of Italy shows that intermittent regional strategies can alleviate the COVID-19 epidemic. Nat. Commun. **11**(1), 1–9 (2020)
9. Fitzpatrick, M., Galvani, A.: Optimizing age-specific vaccination. Science **371**(6532), 890–891 (2021)
10. Ganesh, A., Massoulié, L., Towsley, D.: The effect of network topology on the spread of epidemics. In: 24th Joint IEEE Conference of the Computer and Communications Societies, vol. 2, pp. 1455–1466 (2005)
11. Giordano, G., et al.: Modelling the COVID-19 epidemic and implementation of population-wide interventions in Italy. Nat. Med. **26**, 1–6 (2020)
12. Gourdin, E., Omic, J., Van Mieghem, P.: Optimization of network protection against virus spread. In: 8th International IEEE Workshop on the Design of Reliable Communication Networks, pp. 86–93 (2011)

13. Hønge, B., Hindhede, L., Kaspersen, K.A., et al.: Long-term detection of SARS-CoV-2 antibodies after infection and risk of re-infection. Int. J. Infect. Dis. **116**, 289–292 (2022)
14. Jain, V., Iyengar, K., Garg, R., Vaishya, R.: Elucidating reasons of COVID-19 re-infection and its management strategies. Diabetes Metab. Syndr. Clin. Res. Rev. **15**(3), 1001–1006 (2021)
15. Kermack, W., McKendrick, A.: Contributions to the mathematical theory of epidemics. II.-the problem of endemicity. Proc. R. Soc. London Ser. A **138**(834), 55–83 (1932)
16. Kikuchi, H., Mukaidani, H., Saravanakumar, R., Zhuang, W.: Robust vaccination strategy based on dynamic game for uncertain SIR time-delay model. In: IEEE International Conference on Systems, Man, and Cybernetics, pp. 3427–3432 (2020)
17. Köhler, J., Schwenkel, L., Koch, A., Berberich, J., Pauli, P., Allgöwer, F.: Robust and optimal predictive control of the COVID-19 outbreak. Ann. Rev. Control **51**, 525–539 (2020)
18. Kucharski, A., et al.: Effectiveness of isolation, testing, contact tracing, and physical distancing on reducing transmission of SARS-CoV-2 in different settings: a mathematical modelling study. Lancet. Infect. Dis. **20**(10), 1151–1160 (2020)
19. Liu, Z., Lai, Y., Ye, N.: Propagation and immunization of infection on general networks with both homogeneous and heterogeneous components. Phys. Rev. E **67**(3), 031911 (2003)
20. Liu, Z., Omayrat, M., Stursberg, O.: A study on model-based optimization of vaccination strategies against epidemic virus spread. In: International Conference on Informatics in Control, Automation and Robotics, pp. 630–637 (2021)
21. Liu, Z., Stursberg, O.: On the use of MPC techniques to decide intervention policies against COVID-19. IFAC-Papersonline **54**(14), 476–481 (2021)
22. López, L., Rodó, X.: The end of social confinement and COVID-19 re-emergence risk. Nat. Hum. Behav. **4**(7), 746–755 (2020)
23. Moore, S., Hill, E., Dyson, L., Tildesley, M., Keeling, M.: Modelling optimal vaccination strategy for SARS-CoV-2 in the UK. PLoS Comput. Biol. **17**(5), e1008849 (2021)
24. Ndaïrou, F., Area, I., Nieto, J., Silva, C., Torres, D.: Fractional model of COVID-19 applied to Galicia, Spain and Portugal. Chaos Solitons Fractals **144**, 110652 (2021)
25. World Health Organization: Strategy to achieve global COVID-19 vaccination by mid-2022 (2021)
26. Pastor-Satorras, R., Vespignani, A., et al.: Epidemics and immunization in scale-free networks. In: Handbook of Graphs and Networks. Wiley-VCH, Berlin (2003)
27. Peng, C., Jin, X., Shi, M.: Epidemic threshold and immunization on generalized networks. Physica A **389**(3), 549–560 (2010)
28. Pham, Q., Nguyen, D., Huynh-The, T., Hwang, W., Pathirana, P.: Artificial intelligence (AI) and big data for coronavirus (COVID-19) pandemic: a survey on the state-of-the-arts. IEEE Access **8**, 130820 (2020)
29. Preciado, V., Zargham, M., Enyioha, C., Jadbabaie, A., Pappas, G.: Optimal vaccine allocation to control epidemic outbreaks in arbitrary networks. In: 52nd IEEE Conference on Decision and Control, pp. 7486–7491 (2013)
30. Salvagno, G., Henry, B., Pighi, L., De Nitto, S., Gianfilippi, G., Lippi, G.: The pronounced decline of anti-SARS-CoV-2 spike trimeric IgG and RBD IgG in baseline seronegative individuals six months after BNT162b2 vaccination is consistent with the need for vaccine boosters. Clin. Chem. Lab. Med. (CCLM) **60**(2), e29–e31 (2022)

31. Wan, Y., Roy, S., Saberi, A.: Network design problems for controlling virus spread. In: 46th IEEE Conference on Decision and Control, pp. 3925–3932 (2007)
32. Yang, H., Lombardi J, L.P., Castro, F., Yang, A.: Mathematical modeling of the transmission of SARS-CoV-2-evaluating the impact of isolation in Sao Paulo State (Brazil) and lockdown in Spain associated with protective measures on the epidemic of COVID-19. PLoS One **16**(6), e0252271 (2021)
33. Zanette, D., Kuperman, M.: Effects of immunization in small-world epidemics. Physica A **309**(3–4), 445–452 (2002)

Author Index

© The Editor(s) (if applicable) and The Author(s), under exclusive license
to Springer Nature Switzerland AG 2023
O. Gusikhin et al. (Eds.): ICINCO 2021, LNEE 1006, p. 209, 2023.
https://doi.org/10.1007/978-3-031-26474-0

Printed in the United States
by Baker & Taylor Publisher Services